PRINCIPLES AND PRACTICE OF ELECTROTHERAPY

FOURTH EDITION

Joseph Kahn, Ph.D., P.T.

Clinical Assistant Professor Emeritus
School of Health Technology and Management
State University of New York
Stony Brook, New York

CHURCHILL LIVINGSTONE
A Harcourt Health Sciences Company
New York   Edinburgh   London   Philadelphia   San Francisco

CHURCHILL LIVINGSTONE

A Harcourt Health Sciences Company

The Curtis Center
Independence Square West
Philadelphia, Pennsylvania 19106

Library of Congress Cataloging-in-Publication Data

Kahn, Joseph.
    Principles and practice of electrotherapy / Joseph Kahn. — 4th ed.
        p. ; cm.
    Includes bibliographical references and index.
    ISBN 0-443-06553-5
    1. Electrotherapeutics.   I. Title.
    [DNLM: 1. Electric Stimulation Therapy—methods. WB 495 K123p 2000]
    RM871.K27 2000
    615.8'45—dc21

                                                    99-051477

PRINCIPLES AND PRACTICE OF ELECTROTHERAPY   ISBN 0-443-06553-5

Printed in the United States of America

Last digit is the print number: 9   8   7   6   5   4   3   2   1

About the Author

The author has served as Clinical Assistant and Associate Professor, in addition to Adjunct Lecturer, at the following:

Ithaca College, Ithaca, New York; Albert Einstein Medical Center, Bronx, New York; Daemen College, Amherst, New York; Touro College, Dix Hills, New York; SUNY Downstate Medical Center, Brooklyn, New York

Now retired, he still consults, lectures at seminars, and is a frequent contributor of articles to the literature.

Preface

In the field of electrotherapy, great advances have been made in the sophistication of the equipment used. Computer programs are now integral to the apparatuses with which we work. Still, the astute clinician must be knowledgeable in the complexities and variations of the modalities so that he or she can alter, adjust, and modify treatment plans for individual patients. Although the hardware is designed to establish protocols for all cases, there are always those patients who require different parameters and techniques. It is still up to innovative practitioners to sidestep offered protocols and deliver needed specific procedures when and where applicable. No manufacturer, far removed from the front lines, can establish successful protocols for every patient. Manufacturers, dealers, and sales representatives claim to make things easier for the physical therapists by suggesting protocols (not always clinically valid) and presenting recommended programmed procedures on monitors for the clinician to follow. (Why then spend 5 years learning how to design your own treatment plans?) It is hoped that the theories and practices we learned in school and our many years of clinical experience have served us and our patients well. The new "stuff" is merely window dressing.

I urge physical therapists and others who administer electrotherapy procedures to "think physics" and apply time-tested treatments in their everyday practices, using new and updated equipment for the benefit of the patients in their care.

In recent years I have been corresponding with a Lithuanian physician-physiotherapist, Dr. Raimundas Kibisa. We were able to have sent to him updated electrotherapy equipment donated by persons here and by several generous manufacturers for his research in the use of electrical stimulation to the calf musculature to prevent venous stasis during extensive surgical procedures.

Joseph Kahn, Ph.D., P.T.

Contents

Introduction to Electrotherapy

The concept of electricity as a therapeutic agent is not a recent innovation. The use of electricity for therapeutic purposes has grown in recent years and now includes a wide variety of apparatus, leading one perhaps to the false impression that the concept is novel. Our profession's literature, however, contains many early references to the use of electricity as a therapeutic tool, not only for experimental trial, but also for clinical application. As early as 1757, Dr. Benjamin Franklin wrote in a letter about his administration of "electrical shocks" to his neighbor, Mr. John Pringle, for an obvious frozen shoulder, with good results (Fig. 1-1). Of course, much worldwide experimental and investigative work had preceded Dr. Franklin's clinical approach. Today, some of the standard terminology used in the field of electrotherapy is named for these early workers (e.g., Michael Faraday and Luigi Galvani).

Basic physics, of which electricity is but one branch, also includes the science of light, heat, cold, sound, and mechanics. As physical therapists, we use all of these natural phenomena in our daily practices, adapting them to our specific medical needs. Indeed, if the term *physical therapy* were to be exchanged for another professional title, an appropriate choice would be *applied physics*.

The electromagnetic spectrum, ranging from radio and TV transmission to cosmic and gamma radiations (see Fig. 2-1), is the source for most of the electrotherapy modalities. Receiving radiation from the sun, the human body is able to separate selectively each component according to its anatomic makeup and need and absorb bands in the spectrum for physiologic effects specific to that range of wavelengths or frequencies. Examples of this would be the absorption of infrared radiation (770 to 12,000 nm band) as heat to the 3-mm level of the skin; the absorption of the ultraviolet band (15 to 390 nm) at the 1-mm level, where chemical changes produce sunburn and essential metabolic processes; and, of course, the absorption of the traditional red-orange-yellow-green-blue-indigo-violet (ROYGBIV) rays (770 to 390 nm) we know as the visible spectrum. In the practice of physical therapy we are able to produce these various bands of energy artificially to apply them conveniently in the form of therapeutic modalities.

When applying electromodalities to patients, it is essential for the practitioner to respect the electrical nature of the human body. The vast neurologic network within the body dwarfs NASA when it comes to circuitry, redundant systems, and complex ionic exchanges. An appreciation of the body's electrophysiology and the electrical qualities of the modality of choice is essential to appropriate treatment. As the body functions by electricity, its malfunction can also be measured by using electricity. Many diagnostic procedures are now done electronically, such as the electrocardiogram (ECG), electroencephalogram (EEG), and electromyogram (EMG), as well as those procedures discussed in Chapter 8 of this book on testing.

Although not electrical in nature, ultrasound is often included in discussions on electrotherapy modalities. A soundwave, consisting strictly of a longitudinal form, rather than the typical transverse electromagnetic form, requires a medium through which to travel. However, since many of the physical laws pertaining to radiation also apply to soundwave transmission, and complex electrical components are required to produce the therapeutic soundwaves, ultrasound is traditionally included in the general topic of electrotherapy.

ELECTRICAL TREATMENT
FOR PARALYSIS

To
John Pringle
Craven-Street, Dec. 21, 1757

SIR,

In compliance with your request, I send you the following account of what I can at present recollect relating to the effects of electricity in paralytic cases, which have fallen under my observation.

Some years since, when the news-papers made mention of great cures performed in Italy and Germany, by means of electricity, a number of paralytics, were brought to me from different parts of Pennsylvania, and the neighboring provinces, to be electrised, which I did for them at their request. My method was, to place the patient first in a chair, on an electric stool, and draw a number of large strong sparks from all parts of the affected limb or side. Then I fully charged two six gallon glass jars, each of which had about three square feet of surface coated; and I sent the united shock of these through the affected limb or limbs, repeating the stroke commonly three times each day. The first thing observed, was an immediate greater sensible warmth in the lame limbs that had received the stroke, than in the others; and the next morning the patients usually related, that they had in the night felt a pricking sensation in the flesh of the paralytic limbs; and would sometimes show a number of small red spots, which they supposed were occasioned by those prickings. The limbs,

FIG. 1-1. Letter from Benjamin Franklin. (From the files of the Princeton University Library, courtesy of Dr. Fred Girshick, Department of Nuclear Chemistry.) *(Figure continues.)*

RECENT DEVELOPMENTS

Compared with the many recent advances in medical science, most forms of therapeutic electrotherapy are quite old. Among the newer modalities of electrotherapy is the cold laser. Laser light is produced within the segment of the visible spectrum (632.8 nm), but it does not act as ordinary light and has been found to have unusually favorable effects on human metabolic processes. Ordinary light in this band would be detected and absorbed as a red beam, affecting only the retina and visible cortex of the brain. Laser light, however, has quite a different effect.

Other recent developments in the field of electrotherapy are primarily a furthering of existing modalities with variations in waveforms, amplitudes, polarities, and other parameters. Nothing new has been discovered in the basic electrophysics, but many innovative adaptations and applications have been established due to the findings in studies of human neurophysiology. As more information concerning the operation of the human systems is made available to clinicians, new and effective techniques must be designed to optimize therapeutic applications in physiotherapeutic practice.

Another recent addition to the storehouse of

too, were found more capable of voluntary motion, and seemed to receive strength. A man, for instance, who could not the first day lift the lame hand from off his knee, would the next day raise it four or five inches, the third day higher; and on the fifth day was able, but with a feeble languid motion, to take off his hat. These appearances gave great spirits to the patients, and made them hope a perfect cure; but I do not remember that I ever saw any amendment after the fifth day; which the patients perceiving, and finding the shocks pretty severe, they became discouraged, went home, and in a short time relapsed; so that I never knew any advantage from electricity in palsies that was permanent. And how far the apparent temporary advantage might arise from the exercise in the patients' journey, and coming daily to my house, or from the spirits given by the hope of success, enabling them to exert more strength in moving their limbs, I will not pretend to say.

Perhaps some permanent advantage might have been obtained, if the electric shocks had been accompanied with proper medicine and regimen, under the direction of a skillful physician. It may be, too, that a few great strokes, as given in my method, may not be so proper as many small ones; since, by the account from Scotland of a case, in which two hundred shocks from a phial were given daily, it seems, that a perfect cure has been made. As to any uncommon strength supposed to be in the machine used in that case, I imagine it could have no share in the effect produced; since the strength of the shock from charged glass is in proportion to the quantity of surface of the glass coated; so that my shocks from those large jars must have been much greater than any that could be received from a phial held in the hand. I am, with great respect, Sir,

Your most obedient servant,
B. FRANKLIN.

FIG. 1-1. *(Continued.)*

human science has been the concept of endorphins and the role they play in pain control. The fact that nature's own painkiller, β-endorphin, is produced in the human body is a fascinating phenomenon. The body's production of endorphin is believed to be greatly enhanced by the administration of electrical current to the surface of the body (see Ch. 6). Pain normally triggers the production of this chemical, which provides an analgesic effect. Electrical stimulation, though *not* painful, will also trigger endorphin production and dissemination within the human system. Hence, even standard electrical muscle stimulation, designed to elicit contractions as an exercise modality, may also act as a stimulant for pain control by increasing the amount of endorphin released.

Clinical thermography is now recommended for patient initial evaluation and progress reports as well as documentation and justification. Greater parameter selection and control, as well as the new computerized programming of regimens, complement the treatment arsenal currently available to the clinical physical therapist. The myriad of modalities, brands, models, and specially designed apparatus demand vigilant study and update on the part of the practitioner. Continuing education courses covering these newer aspects are a necessity for today's clinician.

As mentioned throughout this text, the key to successful practice in electrotherapy is a working knowledge of the basic physics of each modality; it is not enough to know which switch to throw or which dial to turn, but why the treatment is selected. The goals of those who desire competence in this area of practice should be the ability to (1) select modalities skillfully; (2) administer effective treatment; (3) adapt appropriately when necessary; and (4) evaluate the results accurately.

2

High-Frequency Currents

Currents with oscillations greater than 10,000 Hz are termed *high frequency*. (Those with oscillations in the 1,000 to 10,000 Hz range are considered as *medium frequency*.) This band in the electromagnetic spectrum has been greatly underutilized in recent years. Recent developments in the high-frequency field have led to increased interest in and production of competitive models for innovative clinicians.

SHORTWAVE DIATHERMY

Shortwave diathermy has been a viable modality in physical therapy for more than 70 years. The concept of producing heat deep within the tissues, beyond the reach of infrared, hot packs, and other forms of superficial heat, is appreciated by clinicians and, where applicable, has proven effective through the years.[1] Originally utilized for musculoskeletal and joint conditions requiring increased circulation[2,3] and other benefits of thermal increase,[4] diathermy has proven an effective modality in the management of nonorthopedic conditions requiring the same benefits. Among these conditions are the chronic obstructive pulmonary diseases (COPD) (e.g., bronchitis),[5,6] urologic diseases (e.g., prostatitis),[7,8] gynecologic conditions[9–14] (e.g., nonspecific pelvic inflammation), and otolaryngologic conditions (e.g., otitis media, sinusitis).[15–18]

In treating the above conditions, a basic law of physics applies: The expansion of gases when heated (Charles' law) aids in expanding the air trapped in the deeper recesses of the body, such as the bronchioles and the sinuses, which, in turn, enhances breathing and drainage that is curtailed by spasmodic and narrowed passageways. The analgesic qualities of heat, the ability to relax skeletal musculature, and the increased capacity for circulatory drainage established shortwave diathermy as a safe, comfortable, and effective modality.[19,20]

PHYSICS
Production of the Shortwave Current

The transformation of standard household output of 120 V at 60 Hz alternating current to well over 500 V at up to 45 MHz is accomplished by an array of electronic components within the chassis of a modern shortwave diathermy unit. Production of this high-frequency current with the eventual "broadcast" to an operational field, much like a radio or TV station sending our waveforms for sensory reception, provides very rapid oscillation or polarity reversal.

The term *shortwave diathermy* is a remnant from the early days of electrotherapy, at which time there were two bands used by physical therapists: *longwave* and *shortwave*. The longwave apparatus often utilized spark-gap devices that interfered with radio transmission and, later, TV, aircraft communications, and so forth. The Federal Communications Commission (FCC) eventually eliminated the *longwave* bands and left the shorter wave lengths for medical use. Costly conversions led to abandonment of existing *longwave* equipment, donations overseas, and contributions to the junk-heap. So, although the frequencies or wavelengths currently allotted to medicine are actually in the high-frequency range, the term *shortwave* has still remained in use.

Ranges

Shortwave diathermy output is regulated by federal controls, which permit only three wavelengths for medical purposes: 7, 11, and 22 m (approximately); or, in terms of frequencies: approximately 45, 27, and 13 MHz.[21] (The wavelength is always the reciprocal of the frequency. The product of frequency × wavelength is always equal to a constant—the speed of light, or 300,000 km/s: $f \times w = c$.) The more common domestic models of shortwave diathermy units are in the 22-m band; some 11-m band models are also available. Most European units, however, offer clinicians the 7-m band apparatus at considerably higher prices.

HEAT GENERATION

The rapid motion of the molecules involved within the field follows yet another basic law of physics dealing with the generation of heat by molecular activity. Closely packed molecules, or dense tissue, naturally become warmer than tissues that are less dense. Differential heating is noted with diathermy between bone, muscle, and fat because of the different densities of these tissues and must be taken into account when treating patients with diathermy, particularly in selecting the electrode type and the placement of electrodes (see below) (Fig. 2-1).

It is apparent from Figure 2-1 that much controversy exists regarding the heating characteristics of diathermy. The first part of the chart states that shortwave diathermy pads heat superficially, yet the diagram shows heating into the bone. The drum technique is certainly the most commonly used today, and this, as shown, provides most heating in the muscle tissue. The term *penetration* when pertaining to shortwave diathermy is misleading. *All tissues within the field* will respond to the high-frequency oscillations, however differentially. Water content and density of each tissue will determine the amount of heat produced. As is shown, fatty tissue becomes heated rapidly with pad electrodes and with microthermy. Assuming pads were to be placed on either side of a limb, the diagram would be incomplete without the opposite pad's effect. In effect, the intervening tissues would be heated subsequent to the localized heating of the fatty

superficial layers. This would not be a factor with the drum technique. Therefore, it is advisable to take extra care to avoid burns from either contact or overheating when using condenser pads.

It is also noted in Figure 2-1 that ultrasound is listed as a heating procedure. I do not ascribe the efficacy of ultrasound to heating effects. Although ultrasound frequencies are indeed *high* (i.e., 1 MHz), they are not of the electromagnetic type and should not be confused with diathermy and microthermy. Further comments on this topic are made in Chapter 4 on ultrasound.

Circuit Resonance or Tuning in the Patient

The shorter the wavelength—or the higher the frequency—the greater the penetration. Technically, shortwave heat treatment is intended to bring the patient into resonance with the unit's output, much the way a radio receiver resonates with the broadcasting studio's output of music or the way a TV set receives the audio/video transmissions of the TV station. In the case of the patient, however, the output is *heat,* deep within the tissues. Diathermy units usually have a specific control, called the *tuner,* designed to bring the patient circuit into resonance with, or in tune with, the main circuit for effective operation.

POWER CONTROLS

The power control is similar to the volume control on the radio or TV; it selects an appropriate heating level. The analogy of the diathermy unit with the radio is a good one and should be kept in mind by all who deal with new, unfamiliar, or innovative models of diathermy units. Some, if not most, newer models offer automatic tuning, eliminating the act of "dialing to find the right station," much like push-button automobile radios. Others offer automatic power control (high, low, etc.) or push-button, preset increments. The automatic tuning operates on a *feedback* system in which the transmitter within the unit is informed, electronically when the patient is in proper resonance, similar to an automatic camera that feeds back data to adjust and determine proper exposure. Power controls, rather than being fully automated, are actually adjusted and/or set by the manufacturer at steplike levels. To operate these controls, the physical therapist

CHOOSING THE BEST HEAT-THERAPY METHOD

Easy-Reading Therapy Chart

Method	°C	Fat	Muscle	Bone	Description
Shortwave: Condenser Field With 2 Opposite Pads					"The Shortwave pads produce superficial heat mainly and are not approved by the Council on Physical Medicine and Rehabilitation." (Note intolerably high temperature rise in fatty tissue.)
Shortwave: Induction Field – Helical Coil/Drum					Shortwave Induction field creates practically no heat in the fatty layer. MOST of its heat is created in the muscle where it can do the most good. Note how close it approaches the bone.
Microwave					Most of the microwave energy is wasted in heating the fatty layer and penetrates muscle only about 1/3 as deep as shortwave. "Microwaves produce an undesirable amount of heat in the subcutaneous fat."
Ultrasound					Ultrasound creates practically no heat in the fatty layer. Therapeutic heat is created in muscle, with maximum rise at the muscle/bone interface and in the bone. Note how ONLY ultrasound can reach a joint!

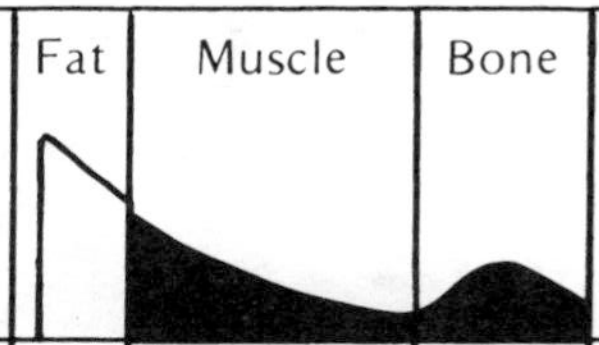

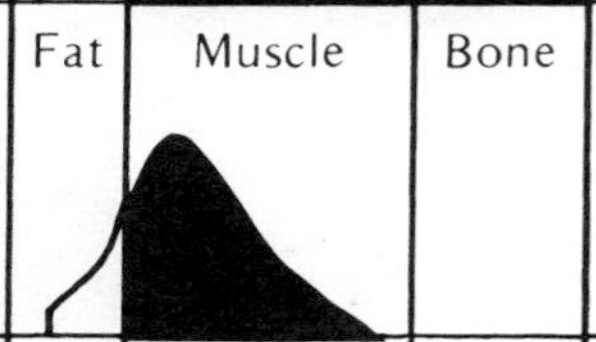

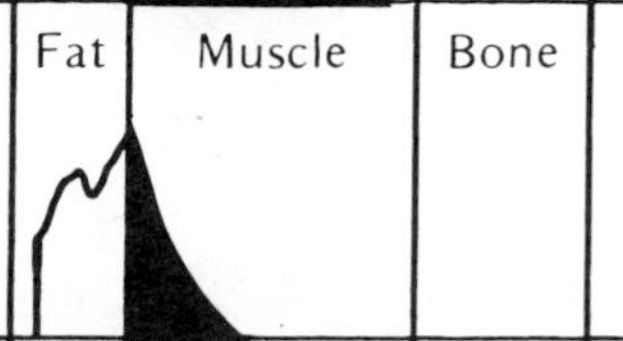

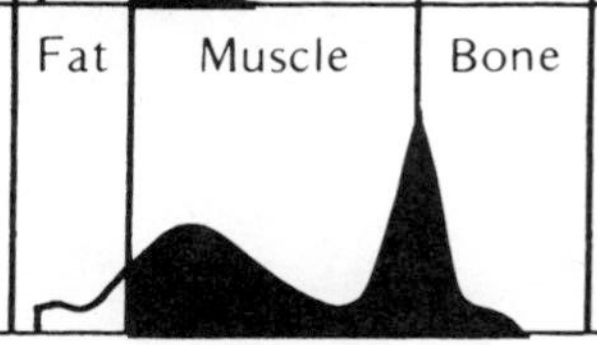

FIG. 2-1. Choosing the best heat therapy method. (Adapted from Paetzold,[31] with permission.)

must refer to the instructions for suggested power settings for various conditions and target distances.

TYPES OF ELECTRODES

Some of the standard texts refer to several distinct types of electrodes used with diathermy apparatus.[22,23] Currently, however, most types fall into two categories, which offer the clinician a wide variety of applicators, depending on the manufacturer, personal preference, and target conditions. Depths of penetration and differential heating of tissues presumably govern the selection. In most instances, however, practitioners use the units available in the facility, utilizing whichever electrodes or applicators are on hand. It is therefore advisable to be familiar and possibly adept with each of them.

Capacitive Electrodes

Capacitive, or condenser, electrodes consist of separate pads or air-spaced drums (Fig. 2-2). An electromagnetic field is created between the two coils, within either the pads or the plastic drum(s). The patient is placed between the electrodes (Fig. 2-3). Heating intensity and depths of penetration are determined by the shape and distance between the electrodes. Because subcutaneous fat is a known insulator, much of the field-energy output is absorbed by the fat, leaving little heating effect for the underlying musculature. This type of electrode, therefore, is most effective when the target tissues are not burdened with overlying layers of fat. Ideally, the field is placed perpendicular to the area to be heated.

Inductive Electrodes

Inductive electrodes are usually hinged-drum types (Fig. 2-4), single drum (Fig. 2-5), or air-spaced induction drums, which can be used singularly. Here the field is truly magnetic in nature, and, according to the natural laws governing magnetism, a field is generated around the coils, emanating outward to include the target area within the field strength of the generating unit. Because this type of field does not become absorbed in subcutaneous fat, penetration to the deeper layers is possible. Absorption is greatest in tissues with high electrolyte content (e.g., blood); however, it has minimal beneficial effect in poorly vascularized tissues such as fat (see Fig. 2-2). Neither do these nonabsorbent tissues dissipate accumulated heat; therefore, they tend to hold heat longer and at higher temperatures in areas where the heat *is* absorbed readily.

FIG. 2-2. Condenser pad electrodes with felt spacers. Full-thickness folded Turkish towels are placed between the spacers and the patient to ensure proper distances and to avoid overheating.

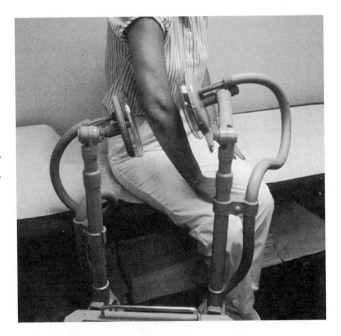

FIG. 2-3. Air-spaced electrodes utilized for the elbow. (Courtesy of Siemens Corp., Addison, Illinois.)

Capacitive vs. Inductive Electrodes

The condenser (capacitive) electrodes may provide greater transsectional heating, concentrating mainly in the subcutaneous fat and adjacent layers. The induction-type electrodes will heat at deeper levels of muscular tissues, bypassing the superficial fat but minimally affecting the tissues with poor blood supplies. The induction type is therefore recommended when heating is needed at deeper levels, passing through subcutaneous fat layers (e.g., quadriceps, shoulder, knee) (Fig. 2-6). Condenser pads may be utilized effectively with superficial targets or in areas where fatty layers are minimal (e.g., posterior chest wall) (Fig. 2-7).

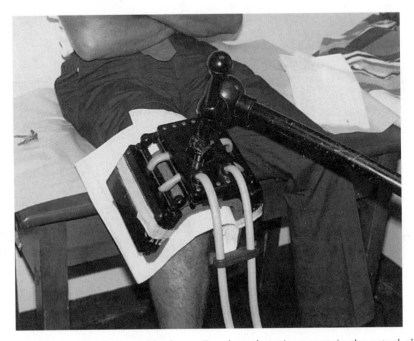

FIG. 2-4. Induction drum technique over the knee. Two lateral sections contain the actual electrodes; the central unit contains the lead connections only. The field is across the target zone.

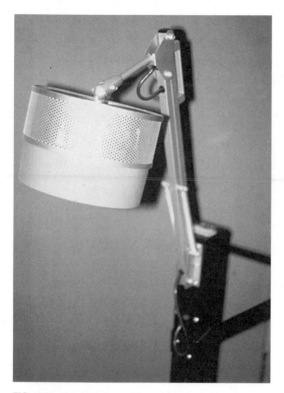

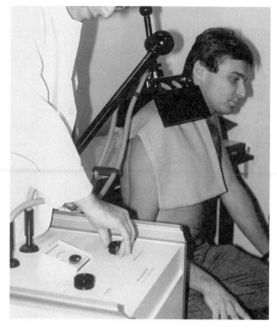

FIG. 2-6. Drum method used over the shoulder; note towel placed under drum and separation between lead wires and the patient's skin. (Courtesy of Birtcher Corp., El Monte, California.)

FIG. 2-5. Single-drum unit, with automatic tuning. (Courtesy of Mettler Co., Anaheim, California.)

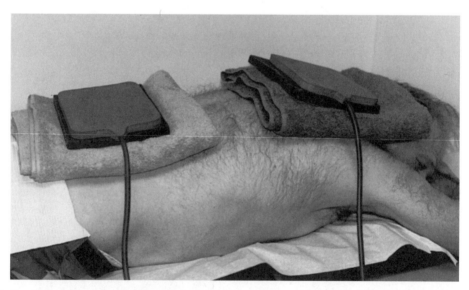

FIG. 2-7. Condenser pad electrodes are seen in place for shallow-depth absorption along the dorsolumbar spine. Lightweight sandbags to maintain position were removed for illustrative purposes. Note that the cables should *not* touch the patient even if additional towel insulation is required.

PULSED DIATHERMY

Although pulsed diathermy is offered by several manufacturers, with claims that nonthermal effects are therapeutically beneficial, no clear evidence now indicates that the on-off phenomenon that minimizes thermal build-up is indicated clinically for any specific condition. With these units, the field generated is pulsed 60 or more times each second, preventing a build-up of heat, since the unit is actually "off" half the time. It is questionable whether oscillating electromagnetic fields alone are therapeutic or whether the heat is the key to effective treatment.

The recent introduction of sophisticated pulsed diathermy apparatus has rejuvenated interest in this modality. It must be noted, however, that many of the claims of superiority over traditional diathermy have yet to be justified. Ease of operation, singular or bipolar drum availability, and multiple frequency selection offer clinicians desired options. On the other side of the coin, however, are the complex controls, large size of the units, and cost.

INDICATIONS

1. For chronic, mild conditions, requiring minimal heating (e.g., myositis, arthritis, pelvic inflammation, otitis media, bronchitis, sinusitis, and prostatitis)
2. As an adjunct modality with other procedures (e.g., following iontophoresis to enhance local absorption, see Ch. 9)

CONTRAINDICATIONS AND PRECAUTIONS

1. Avoid use of shortwave diathermy for acute inflammatory conditions, hemorrhage, or other conditions in which there is fluid build-up, drainage material (leading to increased local absorption and possible burns), tumors, metallic implants, or foreign bodies.
2. Do not use diathermy if the patient has an implanted cardiac pacemaker.
3. Use special care to avoid contact between the patient's skin and the cables from the main unit (Fig. 2-8).

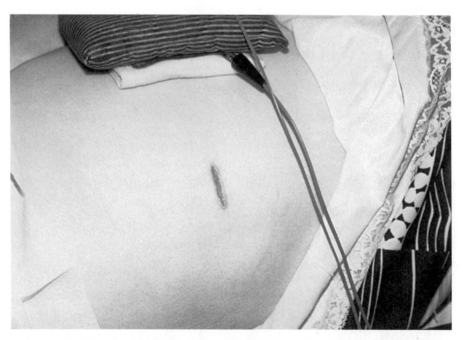

FIG. 2-8. A burn resulting from contact between the patient's skin and a diathermy cable. The patient did not report the discomfort at the facility where this occurred. (In the present facility, treatment, as shown, is electrical stimulation.)

4. Do not allow the patient to come in contact with metallic objects during the treatment (e.g., chairs, lamps, nearby apparatus).
5. Use special care with patients who have arteriosclerosis and diabetes, since these conditions are often accompanied by loss of normal sensation. Susceptibility to burns is common with these patients; healing qualities are poor, and healing time is lengthy.

TREATMENT PROCEDURES

1. Read the manufacturer's operating manuals carefully before administering treatment with an unfamiliar unit. Individual models may differ widely from one another in controls and operation.
2. Warm up units properly, as recommended by the manufacturer.

Electrode Placement

Place electrodes at the correct sites, depending on the condition treated and the type of electrodes used. (See section on electrodes above.)

Capacitive Electrodes

1. If the electrodes are capacitive (pads or air-spaced drums), have the target tissue *between* the two pads or drums (Fig. 2-9).
2. If heating becomes uncomfortable and cannot be lowered by controls on the unit (power), separate the pads or drums further. Wider

spacing reduces the amount of energy absorbed at depths.
3. Place folded Turkish towels on the skin immediately under the electrodes (pads), so that perspiration will be absorbed, minimizing unwanted energy accumulation from the fluid contents of the area.

Induction Electrodes

1. In inductive, hinged-drum models, the central portion of the drum is for mechanical purposes only and contains no diathermy coils.
2. The lateral "wings" contain the coils and create the field when properly positioned by the clinician.
3. It is suggested again that Turkish towels be placed over the target area to absorb perspiration during the treatment.
4. It is not necessary for the drums to be in contact with the towels during the treatment. With pad electrodes, there is usually contact between the pads and the towels, held in place with lightweight sandbags or loosely bound Velcro strapping (Fig. 2-10).

Draping the Patient

1. Do not allow the target area to be hidden by clothing or draping, so that monitoring will be possible.
2. Ensure that there is no overlying clothing or draping, because it can insulate dissipated heat and cause unwanted irritation or burns.

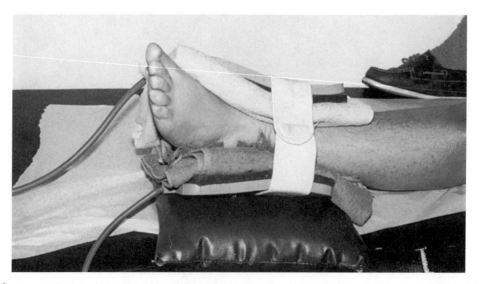

FIG. 2-9. Transarthral technique with condenser pads across the target zone. Note toweling and Velcro security strap.

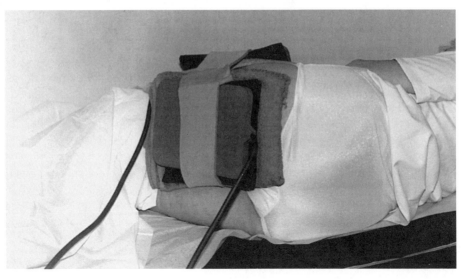

FIG. 2-10. Condenser pads placed across upper thigh in a trochanteric bursitis, held in place with Velcro strap. Patient is more comfortable in the side-lying position than when supine or prone.

3. Have the patient remove tight straps (e.g., bra) and elastic supports (e.g., stockings, girdles, braces), so that circulation is not compromised and heat may be dissipated readily by the sweep of unrestricted circulation.

Setting Initial Power

1. Select the power level based on the operating manual's recommendations, the patient's condition, and the experience of the physical therapist.
2. If in doubt, choose a low intensity, usually at one-third of power, and increase the intensity to provide a mild, comfortable heating sensation once tuning is accomplished.

Tuning the Patient

1. Now bring the patient into resonance (tuned) by manual control until a predetermined heating level is reached.
2. Meter-based or digital readouts provided by the unit may indicate suggested treatment parameters for distance, conditions, or target characteristics.
3. With older equipment, comfortable levels are usually located in the "medium" range on the meter dials.
4. Avoid the "red danger" range on most meters, usually by lowering power output and returning at the adjusted power levels.

Monitoring the Treatment

1. The patient should feel a mild sensation of warmth throughout the treatment.
2. Immediately check any reports by the patient of a "hot" sensation.
3. Again, I emphasize that you should give extra care to patients with impaired sensation (e.g., arteriosclerosis, diabetes, or peripheral vascular insufficiencies).

Automatic Units

In units offering automatic power or automatic (self) tuning, follow instructions given by the manufacturer's operating manual for proper procedures.

DURATION OF TREATMENT

Treatment times may vary between 10 and 30 minutes, depending on the condition and the patient. Shorter sessions are indicated when the condition is mild and the target area is small (e.g., ear, wrist, or sinuses). Larger areas (e.g., low back, posterior chest, or abdomen) may require additional time, up to 30 minutes, to generate sufficient energy in the tissues and be therapeutically effective. Most diathermy treatments can be administered twice weekly for mild, chronic conditions. Daily treatments are not contraindicated, however, when the condition warrants concentrated care (but *not* more heat). Intensities and times should be adjusted under these circumstances.

RECOMMENDED PROCEDURES FOR SPECIFIC CONDITIONS

Listed below are but a few of the many applications of shortwave diathermy. The innovative physical therapist will find applicable targets for the deep heat provided by the modality. Historically, training of physical therapists emphasized treatment according to traditional methods and only to specific conditions. Today's sophisticated clinician, armed with a background in electrophysics and other technical skills demanded by modern modalities, is better prepared to apply diathermy to a wide spectrum of clinical conditions not listed by name, but rather those requiring deep heat, regardless of anatomic or specialty classifications.[24-28]

Sinusitis

1. Have patient assume supine position.
2. Place drum electrode over patient's anterior head and face, with towel insulation (Fig. 2-11).
3. Pad electrodes are not generally used.
4. With patient in seated position, place air-spaced electrodes on either side of the patient's head.
5. With patient seated or supine, use mask electrode and a special sinus mask on face, with second pad on cervical/dorsal spine (Fig. 2-12).
6. Administer treatment for 10 minutes (mild heating sensation).
7. Follow with infrared radiant heat to face.
8. Follow with cervical massage (with patient seated).

Otitis Media

1. Have patient assume supine or side-lying position, with affected ear proximal.
2. Place drum electrode over affected ear with towel insulation (Fig. 2-13).
3. Place air-spaced electrodes on either side of the patient's head (patient seated).
4. Use of pad electrodes is not advised.
5. Administer treatment for 10 minutes (mild heating sensation).

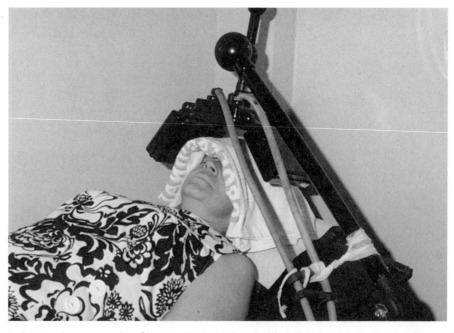

FIG. 2-11. Induction-drum procedure for sinusitis. Note that toweling does not cover the eyes or nostrils so that the patient is more comfortable and breathes easily. The field is across the entire head, including most sinus cavities. (Courtesy of Birtcher Corp., El Monte, California.)

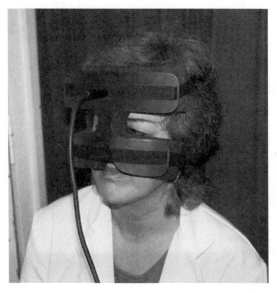

FIG. 2-12. Diathermy sinus mask in place, with the secondary pad placed on the dorsal spine or lumbar region. Special care must be taken to avoid cable to skin contact.

Prostatism

1. Have patient assume prone position; use pelvic pillow support (Fig. 2-14).
2. Place drum electrode over gluteal fold from L5 to perineum.

3. Place pad electrodes over lumbar spine and under anterior abdomen—with additional toweling.
4. Place air-spaced electrodes anterior and posterior across the patient's lower abdomen (patient side-lying).
5. Administer treatment for 20 minutes (mild heating sensation).
6. Provide surged alternating current electrical stimulation to the bilateral sacral nerve roots, at 100 Hz, six surges per minute; 20 minutes is recommended prior to diathermy; if tolerated by the patient, slight visible contractions of the perineum may be elicited.

Bronchitis, COPD, and Other Upper Respiratory Tract Conditions

1. Have patient assume prone position.
2. Place drum electrodes over cervical and dorsal spine (Fig. 2-15).
3. Place pad electrodes over dorsal and lumbar spine (this treatment may also be administered with the patient seated) (see Fig. 2-7).
4. Place air-spaced electrodes anterior and posterior, across the patient's upper thorax (patient side-lying).

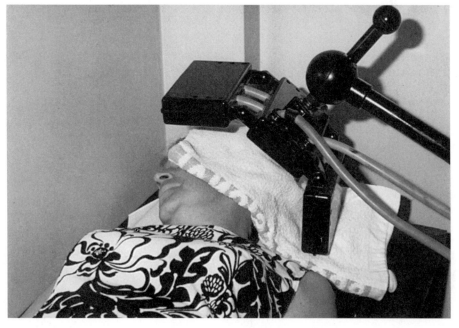

FIG. 2-13. For treatment of otitis media, the patient is supine with the affected ear turned upward. Drums are placed with the field between the two active sections. Eyes and nose are not covered by the towel. (Courtesy of Birtcher Corp., El Monte, California.)

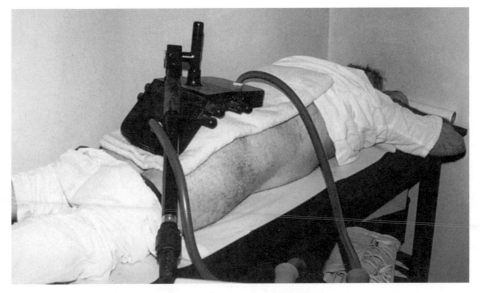

FIG. 2-14. In this treatment of the prostate region, the drums are spread widely, creating a field in the midline area, through the gluteal fold where there is less intervening tissue. Although not shown, a pelvic pillow support is advised. (Courtesy of Burdick Corp., Milton, Wisconsin.)

5. Administer treatment for 20 minutes (mild heating sensation).
6. Ten minutes of infrared radiant heat over the anterior chest is recommended after the above procedure, using a rubefacient consisting of white petrolatum (Vaseline 1 lb), mixed with 0.5 oz methyl salicylate (oil of wintergreen).

Pelvic Inflammatory Disease (PID)[14]

1. Have patient assume supine position, with popliteal support.
2. Place drum electrodes over patient's anterior abdomen, taking special care to insulate with thick towels over the anterior iliac spines (Fig. 2-16).

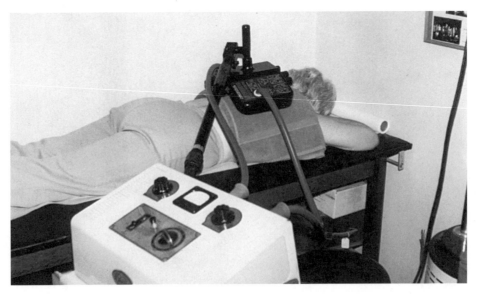

FIG. 2-15. Diathermy technique with induction drums for bronchitis. (Courtesy of Burdick Corp., Milton, Wisconsin.)

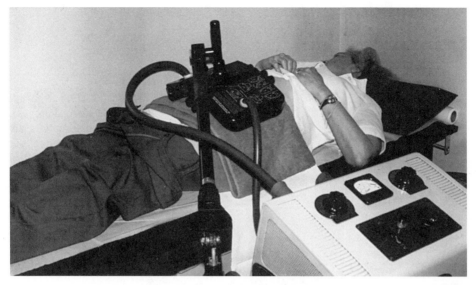

FIG. 2-16. With patient supine, drums are placed so that the entire lower abdomen is in the contact field, with special caution taken in placement over the anterior iliac spines to avoid contact with pads. Popliteal supports are advised. (Courtesy of Burdick Corp., Milton, Wisconsin.)

3. Place pad electrodes under patient's lumbar spine (with additional toweling) and on the anterior abdomen; or have patient side-lying with pads anterior and posterior across lower abdomen, secured with a wide Velcro binder.
4. Place air-spaced electrodes as above, positioned anterior and posterior across the patient's lower abdomen (patient side-lying).
5. Administer treatment for 20 minutes (mild heating sensation).
6. Patients occasionally report vertigo when sitting after the treatment. Allow for this orthostatic adjustment. Follow with a few minutes of cervical massage to hasten recirculation of pooled blood in the abdominal area. Increased perspiration is also common when the abdominal area is heated.

Chondromalacia Patella

1. Have patient assume sitting position, with popliteal support.
2. Place drum electrodes over patient's anterior knee (see Fig. 2-4).
3. Place pad electrodes in transarthral medial and lateral or anterior and posterior positions (Fig. 2-17).
4. Placement of air-spaced electrodes in transarthral medial and lateral positions is advisable.

5. Administer treatment for 15 to 20 minutes (mild heating sensation).
6. Prior to heating, administer 10 minutes of surged alternating current, at 100 Hz, six surges per minute. There will be patient tolerance but no visible contractions, with electrodes placed medial and lateral at knee. No visible contractions are necessary when the transarthral technique is used.
7. Follow diathermy with 10 minutes of infrared radiant heat, using the white petrolatum and methyl salicylate compound described above or Myoflex salicylate ointment.*

Low Back Pain[29]

1. Have patient assume prone position, with pelvic and ankle support.
2. Place drum electrodes over patient's lumbar spine (Fig. 2-18).
3. Place pad electrodes over dorsal and lumbar spine for superficial musculature, or have patient in side-lying position with pads placed anterior and posterior at the lumbar level,

---

*Myoflex (over the counter) analgesic cream, triethanolamine salicylate 10 percent, Warren-Teed Laboratories, Columbus, OH 43215.

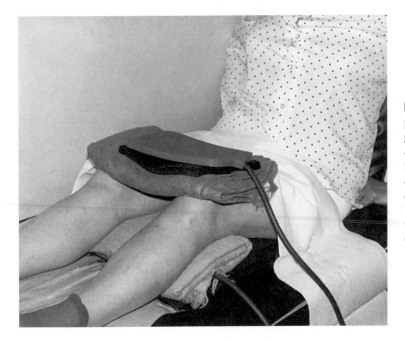

FIG. 2-17. A transarthral approach to bilateral knees in degenerative joint disease, with anterior/posterior placements for convenience. Not shown is a sandbag used to secure the anterior pad during treatment. Note space between extremities to avoid hot spots at skin contacts.

secured with a Velcro binder if deeper musculature is targeted.

4. Place air-spaced electrodes anterior and posterior at the lumbar level (patient side-lying).
5. Administer treatment for 20 to 30 minutes (mild heating sensation).

6. Prior to heating, administer tetanizing alternating current at 100 Hz for 5 minutes, and surged alternating current at 100 Hz for 10 minutes, with electrodes placed bilaterally over the patient's quadratus lumborum and proximal gluteals.

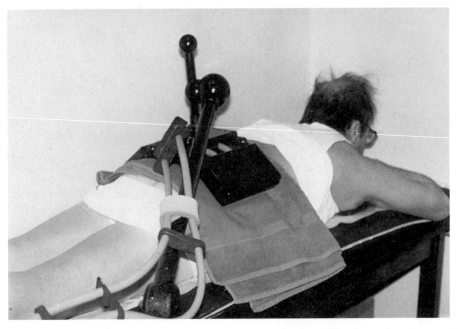

FIG. 2-18. Induction-drum technique for low back conditions. (Courtesy of Birtcher Corp., El Monte, California.)

7. If pain is severe, iontophoresis with an indicated ion (e.g., lidocaine, hydrocortisone, magnesium, mecholyl, salicylate) is suggested *before* heating.
8. Follow with infrared radiant heat with either of the rubefacients mentioned in the previous sections.

One additional application of diathermy is to follow iontophoresis with 5 to 10 minutes of shortwave diathermy. By heating the underlying tissues, more rapid and perhaps deeper penetration of the introduced ions will follow.

MICROTHERMY

Microthermy is similar to shortwave diathermy in concept (i.e., it is also a deep-heating modality). It differs sharply, however, in its mechanics and physics.[30]

FEDERAL REGULATIONS

Medical devices such as diathermy and microthermy units are regulated by the Federal Drug Administration (FDA), as well as by the FCC, in an effort to protect clinicians and the patients under their care. Current regulations limit the users to microthermy units that conform to the standards set for radiation leakage and safety.[21] Before administering treatments, clinicians should be certain that the units they use are within the given guidelines and are calibrated by an authorized bioengineer or otherwise qualified person.

PHYSICS

Frequency and Wavelength

Current FDA regulations limit clinical microthermy frequencies to the 2,450-MHz range (i.e., 10- to 12-cm wavelengths), considerably shorter than those used in shortwave diathermy.[21]

The Magnetron

The extremely high frequencies of microthermy are not obtained from the transformers, oscillators, and similar electronic components familiar to most of us, but rather from a unique device called a *magnetron* (Fig. 2-19). Electrons flow through a "donutlike" iron unit, the magnetron, which is perforated strategically with tiny holes. The electron flow, in a manner similar to air being blown across the top of an empty bottle, creates a frequency response that varies depending on velocity and the driving force of the

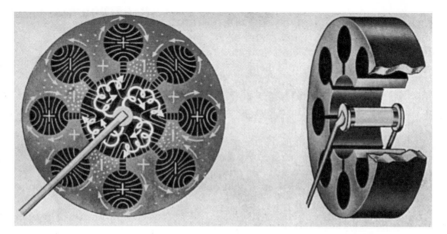

FIG. 2-19. The magnetron oscillator and scheme of electron flow. (From Shriber,[32] with permission.)

electron flow. The harder one blows across the bottle top, the higher the pitch of the produced sound or the higher the frequency. In a similar manner, the high frequency of microthermy is produced within the magnetron and is channeled into the treatment head.

Focusing and Directing the Microwave Energy

The high-frequency energy emitted from the magnetron is directed to the target by the special design of the frequency heads, which shape and focus the field. Because of the high-frequency and power factors, "tuning" by the physical therapist is not required; the therapeutic beam is focused directly at the target area.

Power Levels

Power levels are preselected by the manufacturer and are printed clearly on each director head for reference. Clinicians are advised to follow these directions carefully.

Target Distance

Target-distance relationships are obviously linked to an accepted law in physics: The intensity of radiation is inversely proportional to the square of the distance from the source—the inverse square law. This law states that as one goes further from the radiation source, the intensity of the radiation decreases as the distance is squared, or increases as the square of the distance becomes shorter: stated mathematically, $I = 1/d^2$.

Penetration and Absorption

Although penetration is theoretically increased with the higher frequencies, absorption still depends on the target densities and intervening tissues. Fat, muscle, bone, and connective tissue—including blood—absorb microwave radiation at differing rates; thus, the full clinical effects of microwave radiation, like diathermy, depend on the type of tissues involved. Researchers differ in their assessment of absorption rates of various tissues. In reviewing their findings, one must take into account the motivation and clinical experience of each source (see Fig. 2-1). The heat produced by microthermal radiation is more concentrated than is that of shortwave dia-thermy, albeit not as penetrating because of differential absorption of the radiated energy by intervening tissues. Familiarity with the numerous data concerning absorption by differential tissues is advised for those contemplating administering this modality.

There is still controversy regarding the differential heating characteristics between diathermy and microthermy, especially when fatty tissue is involved. Traditionalists favor the heat produced by diathermy to be greater in muscle tissue because of the high water content, as opposed to fatty layers, which seem to be heated more by the microthermy. It should be noted that the heat produced is more a factor of *absorption* of energy than of linear *penetration*.

PHYSIOLOGY

The thermodynamic responses in the target tissues are similar to those noted in the section on shortwave diathermy; however, the responses differ in absorptive qualities and quantities as previously mentioned.

INDICATIONS

The clinical indications for microthermy are similar to those for shortwave diathermy. In cases in which more concentrated or localized heating is needed, microthermy may be desired over traditional diathermy.

CONTRAINDICATIONS AND PRECAUTIONS

1. Do not use microthermy in the presence of pacemakers.
2. Use special caution in the genital areas; embryonic tissue is hypersensitive to radiation of any type.
3. Avoid radiation to the eyes.

TREATMENT PROCEDURES
Preparing the Patient

1. Clean skin at target region, with overlying clothing removed.

2. Have patient seated or lying down for comfort and accessibility to target.
3. No towel insulation is required with microthermy.

Selecting the Treatment Director (Head)

Treatment head selection will depend on the shape of the field desired and the target's anatomic configuration. In general, directors are found in three designs: small circular, large circular, and rectangular. Each offers a shaped field of energy radiation matching the director's reflector (e.g., small circular target zone, large circular target zone, and elongated rectangular target zone). The actual distance from the director to the skin will determine the extent of each treatment zone (Fig. 2-20). (Using an ordinary fluorescent light tube, the clinician can demonstrate the extent of the radiation being administered by passing the tube between the skin and the director, describing the concentrated area of intensity as compared with the larger area that is treated in a similar procedure with shortwave diathermy.)

Selecting Power Levels

Target distance will determine power levels according to the specific instructions on each director head. A distance regulator in the form of a

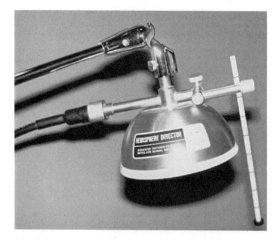

FIG. 2-20. Large hemispheric director (head of microtherm unit, showing the plastic spacer-ruler used to determine distance/dosage factors. A smaller hemispheric director is available, as is a rectangular unit. Each has a spacer-ruler attachment. (Courtesy of Birtcher Corp., El Monte, California.)

plastic ruler is usually part of the microtherm device and allows the clinician accurate measurement of skin distances and subsequent power settings (see Fig. 2-20). The closer the director to the skin, the lower the power levels; the further away, the higher the settings must be. I recommend median distance/power adjustments so that changes may be made in either direction if necessary (i.e., too hot or not hot enough).

Warm-Up Time for Unit Before Application

Most units require a lengthy warm-up period to allow the magnetron to develop enough efficiency. The prescribed warm-up time will usually be given in the manufacturer's operation manual.

Heat Sensation

Patients will report highly localized sensations of heat under the director almost immediately. A gentle warmth is recommended; reports of "hot" spots should be attended to immediately, either by reducing power or increasing distance.

Treatment Time

Microthermy treatments generally require between 10 and 20 minutes, with the smaller target areas (e.g., face, wrist, elbow, ankle) taking less time, whereas larger areas (e.g., low back, abdomen, shoulder) require longer treatment sessions.

REFERENCES

1. Wilensky J, Aronoff G: Therapeutic use of heat. R_x: Home Care, July 1985, p. 70
2. Abramson D, Bell Y, Rejal H et al: Effect of shortwave diathermy on blood flow. Am J Phys Med 39:87, 1960
3. Abramson D, Beaconsfield P: Shortwave diathermy and blood flow. Arch Phys Med Rehab 38:369, 1957
4. Binder A, Parr G, Hazleman B, Fitton-Jackson S: Electromagnetic therapy for rotator cuff tendinitis. Lancet 1:695, 1984
5. Bierman WS, Licht S: Physical Medicine in General Practice. 3rd Ed. Hoeber, New York, 1952
6. Wadsworth H, Chanmugam A: Electrophysical Agents in Physical Therapy. Science Press, Marrickville, Australia, 1980

7. Segura JW, Opitz JL, Greene LF: Prostatosis, prostatitis or pelvic floor myalgea. J Urol 122:168, 1979

8. Woodward WW: Inductotherm in anuria. Lancet, 1957, p. 273

9. Jahier R, Tillier R: Case of amenorrhea treated with high frequency current. J Radiol Electrol (France) 28:55, 1947

10. Upton J, Benson G: Results with local heating in PID. JAMA 121:38, 1943

11. Bengston B: Sustained internal radiant heat in lesions of the pelvis. Arch Phys Ther 24:2632, 1943

12. Kottke F: Shortwave diathermy in pelvic heating. Arch Phys Med Rehab 36:36, 1955

13. Gronroos M, Liukko P: Treatment of vulval epithelial lesions by pulsed high frequency therapy. AC Obstet/Gynecol Scand 58:157, 1979. Abstracted in APTA/OB-GYN Bull 4:1, 1980

14. Balogun JA, Okonofua F: Management of Chronic Pelvic Inflammatory Disease with Shortwave Diathermy. Physical Therapy 66:1541–1545, 1988.

15. Bierman W, Licht S: Physical Medicine in General Practice. 3rd Ed. Hoeber, New York, 1952, p. 719

16. Wadsworth H, Chanmugam A: Electrophysical Agents in Physical Therapy. Science Press, Marrickville, Australia, 1980, p. 71

17. Anonymous: Shortwave diathermy for otitis (progress report). APTA June 1984, p. 6

18. Kahn J: Clinical Electrotherapy. 4th Ed. J. Kahn, Syosset, NY, 1985, p. 95

19. Kahn J: Shortwave Diathermy—Underestimated Potential. Whirlpool (APTA/PPS), 1978, p. 26

20. Kahn J: Shortwave diathermy—dynamite or dinosaur? PT Forum 3:9, 1984

21. Anonymous: Therapeutic Microwave and Shortwave Diathermy. US Department of Health and Human Services, Rockville, MD, December 1984

22. Anonymous: Shortwave Diathermy Units, Health Devices. Vol. 8. Emergency Care Research Institute, Plymouth, VA, June 1979, p. 175

23. Anonymous: Shortwave Diathermy: Assessment Report Series. Vol. 1, No. 16. US Department of Health and Human Services, Rockville, MD, 1981

24. Quinn P (compiler): Bibliography on Diathermy. Information Central, APTA, Alexandria, VA, 1982

25. Boyle JR, Smart B: Stimulation of bone growth by shortwave diathermy. J Bone Joint Surg 45A1, 1963

26. Amundsen H: Thermotherapy and cryotherapy—effects on joint regeneration in rheumatoid arthritis. Physiother Can 31:258, 1979

27. Taylor GA: Treating elephants with short-wave diathermy. Physiotherapy 56:62, 1970

28. Daels J: Induction heating of abdomen during labor—Ghent, Belgium. As reported in Medical World News, June 7, 1974, p. 129

29. Guy's Hospital, London (team): Shortwave diathermy in low back pain. Lancet 1:1258, 1985

30. Anonymous: Microwave Diathermy. 3rd Ed. Burdick Corp., Milton, WI, 1972

31. Paetzold J: Physical laws regarding the distribution of energy for various high frequency methods applied in heat therapy. Ultrasound Med Biol 2:58, 1956

32. Shriber WJ: A Manual of Electrotherapy. 4th Ed. Lea & Febiger, Philadelphia, 1981, p. 208

3

Radiation

PHYSICS

The Electromagnetic Spectrum

Electromagnetic radiation covers a broad spectrum of wavelengths and frequencies, ranging from the longwave-frequency radio-TV and shortwave diathermy bands, through the familiar visible spectrum, to the shortwave–high-frequency bands of ultraviolet, x-rays, and the ever present cosmic radiation.[1] The visible spectrum of red-orange-yellow-green-blue-indigo-violet (ROYGBIV) serves as a fulcrum for radiation modalities, flanked by infrared on one end and ultraviolet on the opposite (Fig. 3-1). Older texts refer to infrared and ultraviolet as the *near* and *far* bands, respectively. These references deal only with the relative position of the two bands within the visible band. Wavelengths closer to the visible end are termed *near* bands, whereas those further from the visible end are termed *far*.

Wavelengths and Frequencies

The formula for the relationship between wavelength and frequency is a universal law in physics that should be committed to memory for future references in radiation studies (Tables 3-1 and 3-2).

$$\text{wavelength} \times \text{frequency} = \text{speed of light, or}$$
$$w \times f = 300,000 \text{ km/s}$$

The range of the visible spectrum is from 770 nm *(red)* through 390 nm *(violet)*. Infrared falls beyond the 770-nm band, and ultraviolet falls beyond the 390-nm band. The cold laser, however, falls directly within the red band, specifically at 632.8 nm. Thus, the cold laser, with which clinicians work today, is a visible wavelength, with physiologic and therapeutic manifestations drastically different from those of ordinary (multiwavelength) light (see below).

Before specific spectrum bands are described, it should be mentioned that all electromagnetic radiation is believed to be similar in form (i.e., waves and particles in motion), differing only in wavelength/frequencies and the associated physiologic effects of each band.

Penetration and Absorption

All electromagnetic radiation is transmitted without the need of a medium or conductor, istransverse in waveform, and may be absorbed, reflected, and/or refracted by various substances and tissues. The electromagnetic radiation received on Earth travels through 93 million miles (150 million km) of space vacuum.

Penetration depends on the wavelength and frequency of the radiation, the angle at which it strikes the surface of the target tissues, and the intensity of the emitting source. Clinically, however, we think more in terms of absorption, since the effective qualities of specific radiation are to be manifested within the tissues where absorption takes place, not necessarily the point or depth of penetration. For example, the ultraviolet radiation we receive from the sun is absorbed, for the most part, at a level of 1 mm, whereas the infrared is absorbed at a level of 3 mm, yet both strike the surface of our body at the same point! The physical therapist, therefore, must select the target tissue with care before applying these modalities clinically. Depth of penetration is often confused with the level of absorption, but the absorption level takes precedence.

The Inverse Square Law

All radiation is subject to the inverse square law, which states that the radiation intensity is

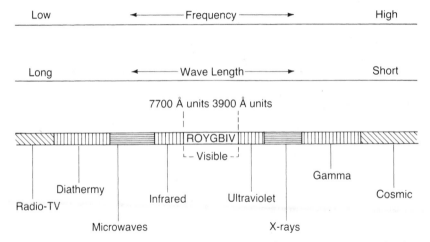

FIG. 3-1. The electromagnetic radiation spectrum. $f \times w = C$; $C = 300,000$ km/s.

inversely proportional to the square of the distance from source to target (Fig. 3-2). In other words, the closer you get, the stronger the radiation; however, the strength increase is at the square of the distance. Conversely, there is a consequent decrease in the intensity as the distance is increased, also by the square of the distance involved. In mathematical terms:

$$I = 1/d^2$$

For example, if a lamp 24 inches (60 cm) away from the skin is moved to a distance of 48 inches (120 cm), the intensity of the radiation is cut to 25 percent of the original strength.

The Cosine Law

Another premise of basic physics pertaining to all radiation is the cosine law, which states that the intensity of radiation varies in relation to the cosine of the angle of incidence (Fig. 3-3): the greater the angle at which the radiation strikes the skin, the less the intensity. In mathematical terms:

$$I = \cos A$$

TABLE 3-1. The Electromagnetic Spectrum

Discoverer	Date	Wavelength (in Air) (nm)	Designation	Frequency per Second	Medical Use
Millikan	1921	0.0001	Cosmic	3×10^{21}	Not known
The Curies } Becquerel }	1898	0.01	Gamma	3×10^{24}	Radium therapy
Roentgen	1895	50	X-ray	5.9×10^{15}	Diagnosis and therapy
Ritter	1801	300	Ultraviolet	7.5×10^{14}	Diagnosis; vitamin D; germicidal; and therapy
Newton	1704	400–800 mm	Visible	3.7×10^{14}	Not known
Herschel	1839	0.001–0.1	Infrared	3×10^{11}	Superficial tissue radiant heating
Hertz	1886	0.5–100 m	Microwave	1×10^{10}	Diathermy heating
Maxwell	1865	10–30	Shortwave	3×10^{5}	Diathermy heating, surgical diathermy
		10–30	Television		Audiomental stimulation
		300	Radio	3×10^{4}	
Faraday	1831	5×10^{10}	Electric power	60	Electrotherapy; neural and muscular stimulation

(From Schriber,[1] with permission.)

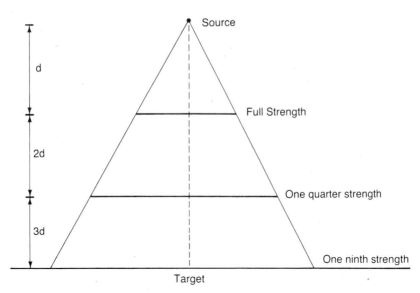

FIG. 3-2. The inverse square law. The intensity of radiation varies inversely as the square of the distance. At twice the distance (2*d*), the intensity is only one quarter of that at *(d)*; at three times the distance (3*d*), the intensity is one ninth of that at *(d)*.

For example, if a lamp is directly over the target zone (i.e., at an angle of 90°), the strength is maximum. As the lamp is angled away from vertical, the intensity decreases in the same ratio as the cosine of the new angle incidence. When the angle becomes too obtuse, reflection usually diminishes the penetration and absorption to zero.

Refraction

Refraction refers to the change in direction of the radiation vector as it passes through different media with variable indices of refraction. Refraction also inhibits absorption in most instances; however, with the cold laser, the wide dispersion and refractive phenomena are favorable, as is shown later.

TABLE 3-2. Wavelengths of Radiations

Radiation	Wavelength
Very long electric waves	5,000,000 m
Radio waves	30,000–1 m
Commercial broadcast	600–200 m
Amateur broadcast	175–20 m
International broadcast	50–15 m
High-frequency (long-wave diathermy)	300 m
High-frequency (short-wave diathermy)	303 m
Microwave	0.5–100 cm
Infrared rays	100,000–770 nm
Visible rays	770–390 nm
Ultraviolet rays	390–13.6 nm
Roentgen rays	13.6–0.14 nm
Gamma rays	0.14–0.001 nm

(From Schriber,[1] with permission.)

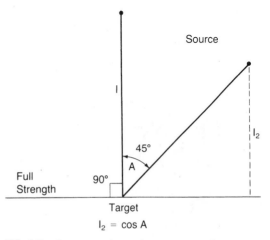

$$I_2 = \cos A$$

FIG. 3-3. The cosine law. The intensity of radiation varies as the cosine of the angle of incidence. Intensity (represented by the dotted line I_2) diminishes as compared with solid line I.

INFRARED

Radiation in the electromagnetic spectrum will be absorbed by the body as heat if the wavelengths fall within the visible band from 770 nm to the shortwave bands at the 45-MHz range. Infrared specifically will be absorbed in the range of 770 to approximately 12,000 nm. Absorption will take place within the top 3 mm of tissues.

PHYSICS

Glowing substances such as those in household toaster elements and ordinary light bulbs emit infrared radiation, which, although invisible, produces profound effects on heat-sensitive tissues. All physical laws pertaining to radiation (and listed above) apply to infrared. These laws of physics relate to the directness and distance of the radiation source from the target and are of great clinical importance.

When using devices emitting infrared radiation, the physical therapist must constantly be aware of the target distance and the angulation of the projected radiation. Reduction of the intensity of the heat produced may be accomplished clinically by (1) increasing the distance from the patient's skin or (2) increasing the angle at which the radiation strikes the skin.

SOURCES OF INFRARED

Both radiation and infrared sources are commonly utilized by physical therapists to adminis-ter superficial heat (Fig. 3-4). Lamps of different types, varying size and wavelength, are found universally in hospitals, clinics, and private office facilities. Used generally to maintain warmth over a prescribed anatomic zone during treatments, or occasionally as the treatment, infrared radiation provides a beneficial and comfortable warmth.

Luminous Units

Luminous units are those that emit infrared radiation derived from glowing or incandescent sources such as hot wires and various types of bulbs (Fig. 3-5).

Nonluminous Units

Nonluminous units are those that produce infrared radiation derived from a nonglowing source, such as household radiators, moist heat packs, bimetallic burners, and chemical heat packs.

PHYSIOLOGY

The numerous benefits of infrared radiation include increased circulation, a sedative effect on superficial nerve endings, lowering of blood pressure, increased respiration and perspiration, loss of salt, water, urea, and other nitrogenous substances, and a general increase in reticuloendothelial activity. Luminous and nonluminous types of lamps differ physiologically, but only in

FIG. 3-4. Typical infrared sources. A 250-W ruby-red bulb (luminous) and a bimetallic ring burner (nonluminous), both designed for use with standard lamp sockets.

FIG. 3-5. Basic infrared lamp, with 250-W ruby-red bulb in a parabolic reflector. (Courtesy of Brandt Corp., Bronx, New York.)

absorptive qualities. There is very little clinical evidence of the efficacy of one over the other.[2] It has been suggested that the nonluminous units apparently "feel hotter" than do the luminous ones at equal distances and power levels, due to the increased absorption of the longer wavelengths by the top layers of the skin. Penetration, however, like absorption, is a function of wavelength and will vary with each unit.[3]

INDICATIONS

Treatment with infrared is indicated for the following conditions.

1. Mild or chronic pain[4]
2. Various manifestations of inflammation (e.g., arthritis)[5]
3. Circulatory conditions (with caution, since the dissipation of heat depends on circulating blood)[6]

4. Adjunct therapy with iontophoresis, electrical stimulation, ultrasound, mobilization, massage, and exercises

CONTRAINDICATIONS AND PRECAUTIONS

Do not use infrared radiation with patients whose sensory levels have been compromised or with whom special precautions and monitoring must be instituted. In many instances, other types of ionizing radiation preclude infrared therapy.

Although there is no danger to the eyes, it is wise to avoid prolonged exposure of the retina to the burning sensation of infrared radiation; use moistened gauze pads over the patient's eyes when the patient's face receives long exposures.

TREATMENT PROCEDURES

Luminous Lamps

1. In general, place luminous lamps directly over target areas at a distance of 24 to 36 inches (60 to 90 cm).
2. Treatments range in duration from 10 to 30 minutes, depending on the target. Face, head, and neck are usually treated for shorter durations (i.e., 10 to 15 minutes), whereas the lower back may be irradiated for more than 20 to 30 minutes.
3. Some filtration is recommended with rubefacients such as creams and ointments that are designed to offer superficial anesthesia and/or hyperemia.

Nonluminous Lamps

1. Place nonluminous lamps, which often feel hotter than their luminous counterparts, at slightly greater distances (i.e., 30 to 42 inches [75 to 105 cm]) and at an angle of about 45°.
2. Treatment times are approximately the same as for luminous lamps.

ULTRAVIOLET

Radiation in the 15- to 390-nm band is absorbed by the body at a depth of approximately 1 mm.

The effects are mainly chemical, in contrast to the pure heating effect of infrared radiation

TABLE 3-3. Comparison of Infrared and Ultraviolet Radiation

Infrared	Ultraviolet
Physical effect	Chemical effect
Absorbed as heat	No heat
Absorbed at 3 mm	Absorbed at 1 mm
Luminous and nonluminous sources	Luminous sources
Immediate erythema	Delayed erythema
Lasts 20–30 minutes	Lasts several days
Mottled	Sharply defined
Dark, reddish	Light pink, homogenous
Occasional tolerance	Progressive
Tolerance	Peeling

(Table 3-3). Ultraviolet causes changes in various chemical processes in the body, including the increased production of steroids.[7] This type of radiation has been utilized for many years in the management of psoriasis and other skin conditions.[8] It is, however, a useful modality with other diseases and conditions, among which are alopecia,[9] dermatitis,[10] furunculosis,[11] herpes zoster,[12] impetigo,[11] lupus vulgaris,[11] pityriasis rosea,[12] osteomyelitis,[11] ringworm,[11] and ulcerations.[11]

More than just a "sunlamp," ultraviolet treatments provide therapeutic dosages of radiation required for various metabolic processes, healing, and general health maintenance. Professional equipment used by physical therapists should not be compared or confused with commercial sunlamps for tanning and cosmetic purposes. The intensities, wavelengths, and techniques of ultraviolet equipment differ as widely as the purposes.

PHYSICS

Mercury vapor has been the accepted source of ultraviolet radiation in therapeutic lamps for many years; carbon-arc lamps were used for decades, however, and can still be found in clinics throughout the world. High-voltage generators provide the energy necessary to vaporize droplets of mercury housed in special quartz-glass tubing. These droplets are placed within reflecting surfaces designed to project the radiation outward in a prescribed field.

Absorption

All ultraviolet is absorbed by the skin to depths of less than 1 mm (except 360 nm, as discussed below). Radiation is usually filtered out by dust, dirt, smoke, leaded glass, and various chemicals.

PHYSIOLOGY
Near Bands

Each lamp is designed specifically to provide certain wavelengths for medical purposes. The *near* bands (i.e., those close to the visible spectrum) are used mainly for antirachitic effects and pigmentation, and with a specific band at 360 nm for the PUVA (psoralen plus ultraviolet A) technique. The remaining bands are commonly termed the *B* bands. The 360-nm band is not normally absorbed by the skin, necessitating ingestion of a photosensitizing agent (e.g., psoralen) to enhance absorption. Sometimes referred to as *black light,* 360 nm is useful in the management of psoriasis (see Indications below).

Far Bands

The *far* bands, or those removed from the visible spectrum, are utilized for the most part for erythemal dosages. These wavelengths are generally classified in the range of 180 to 290 nm, whereas the near bands extend from 280 to 400 nm.

Lamps

Commercial lamps today are generally divided into two types: hot quartz and cold quartz. Hot quartz provides ultraviolet radiation in the combined bands and in A and B regions, whereas cold quartz generally emits in the far bands for erythemal effects (Fig. 3-6).

Wavelengths

Physiologic effects vary with wavelengths. Some overlap occurs, and limitations are relative, especially since most operational lamps emit multiple wavelengths.

Range (nm)	Application	Effect
250–297	Erythema	Red
300–400	Pigmentation	Tanning
240–300	Antirachitic	Metabolic
360–390	PUVA	Psoriasis

Beneficial Effects[9]

1. Vasodilation
2. Activation of steroids
3. Vasomotor stimulation
4. Analgesic to nerve endings
5. Increased muscle tone
6. Metabolic stimulant
7. Sterilization qualities
8. Bactericidal and antirachitic characteristics
9. Psychological benefits[1,7]

Erythema

In contrast to the transient hyperemia that occurs with infrared, ultraviolet produces erythema with generalized, systemic effects.

Effects of Erythema

1. Pro-vitamin D production with low dosages
2. Germicidal effects with increased dosages
3. With extremely high dosages, blistering and destructive effects, used in the treatment of erysipelas and certain types of open ulcerations

Factors Determining the Degree of Erythema

1. Individual patient sensitivity
2. Intensity of radiation source
3. Distance of lamp to target
4. Angle of incidence of radiation at skin
5. Duration of exposure
6. Skin texture: Flexor surfaces are more sensitive than extensor surfaces; blondes and redheads are more sensitive; and children receive half-dosages.

INDICATIONS

Conditions warranting ultraviolet include[8–13]

1. Dermatologic conditions
2. Calcium and phosphorus diseases

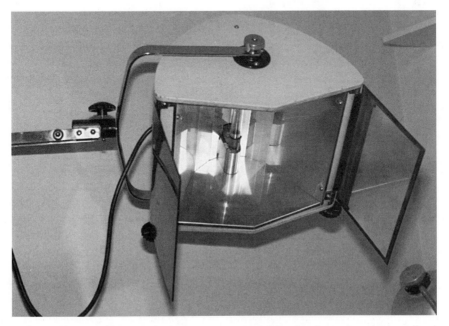

FIG. 3-6. Traditional ultraviolet lamp, with mercury-vapor-emitting tube shown here in the reflecting housing, with "barn doors" in the open position. (Courtesy of Hanovia Corp., Newark, New Jersey.)

3. Nonpulmonary tuberculosis
4. Local ulcerations
5. Upper respiratory condition management (e.g., the common cold)[1]

CONTRAINDICATIONS

Contraindications to the use of ultraviolet include[9]

1. Pulmonary tuberculosis[7]
2. Severe cardiac disturbances
3. Lupus erythematosus
4. Severe diabetes
5. Abnormally acute skin irritations
6. Known photosensitivity
7. Photosensitizing medications

PRECAUTIONS

1. The cornea absorbs all wavelengths beyond 295 nm. Cover the patient's eyes at all times during exposure.
2. Time all treatments carefully—to the second.
3. Never leave the room during an exposure.
4. Wear protective goggles when administering treatments. Photochromic eyeglasses will protect the physical therapist only from frontal radiation, not from the reflected radiation of the lateral aspects.
5. Be aware of the intensity of the reflected radiation from walls, sheets, and equipment, especially when multiple treatments are administered during a single day.

TREATMENT PROCEDURES
Determining the Minimal Erythemal Dosage

Determination of the minimal erythemal dosage (MED) should precede all ultraviolet treatment. It is the preferred test procedure and is used to determine the minimal dosage required to obtain a slight "pinking" of the target skin.

1. Utilize a large piece of cardboard for testing MED.
2. Cut a hole 2 inches (5 cm) in diameter for the test area.

3. Move this hole along the surface at 3-inch (7.5-cm) intervals, offering at least four exposures for later comparison.
4. Drape the rest of the patient's skin so that only the skin in the test opening of the cardboard is exposed to the ultraviolet.
5. Expose small areas on the patient's skin, usually the volar surface of the forearm, with increments of 15 seconds each.
6. Re-examine the skin in 24 hours for comparison.
7. When MED determination is impractical due to travel, distance, logistics, or time factors, commence treatment at minimal exposures (i.e., 20 to 30 seconds) and base calculations on comparative exposure experience with similar skin types (Fig. 3-7).

Dosages

1. Once the MED is known, treatment may be started with increments of 15 seconds at each successive treatment until a maximum is reached, usually at the 2-minute mark for most lamps currently available.
2. Burning and peeling are signs of too much radiation for an MED, and radiation should be reduced.
3. If treatments are effective, peeling and "sunburn" should not be evident.
4. The patient may become slightly tanned after many exposures, but for the best therapeutic effects, this should be a gradual phenomenon.

Distance

1. In general, use a target distance to skin from lamp of 24 to 30 inches (60 to 75 cm), assuming an angle of incidence of 90°.
2. With additional angulation at a distance of 24 to 30 inches (60 to 75 cm) the intensity will be compromised because of the cosine law, mandating slightly increased dosages (e.g., 10 percent) for effective results.
3. If the target distance is less than 24 to 30 inches (60 to 75 cm), reduce dosages accordingly.
4. If the distance is greater than 30 inches (75 cm), institute increased dosages.

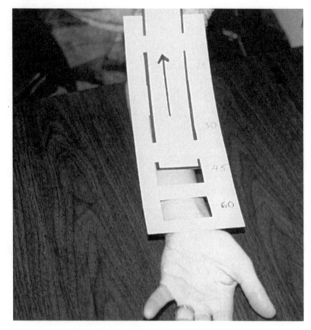

FIG. 3-7. Minimal erythemal dosage (MED) device. Each square opening is exposed for periods with 15-second increments, beginning with a duration of 15 seconds by means of the sliding cover.

RECOMMENDED TECHNIQUES FOR SKIN CONDITIONS

Most dermatologists recommend total body radiation when treating systemic diseases such as psoriasis, even though the skin manifestations may be highly localized. This preference is based on the systemic nature of many dermatologic conditions, in addition to the dermal symptoms.

When total body exposure is utilized, hypersensitive areas of the body should be protected from overexposure by moist gauze pads. Such areas include the male genitals, female nipple areas, and anatomic high points such as the nose, abdomen, and gluteals, depending on the position of the patient (Fig. 3-8).

Filters

1. Use ointments, salves, or plain white petrolatum over the target tissues, to serve as differential filters for effective wavelengths. These

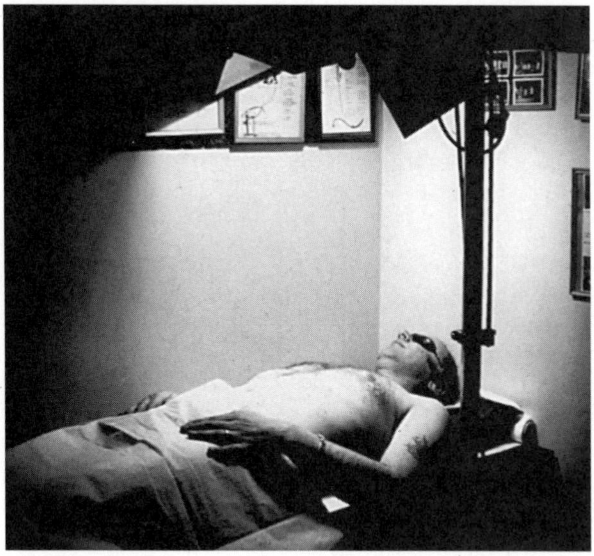

FIG. 3-8. Total body radiation technique for psoriasis. Here the patient is radiated in quarter sections: upper and lower anterior and upper and lower posterior. Target distance is approximately 36 inches (90 cm), directly over the area to be radiated.

filters block out the shorter wavelengths that cause sunburn and permit the longer wavelengths to be transmitted for their therapeutic effect.

2. Still used today, the Goeckermann procedure for psoriasis uses crude coal tar ointment to narrow the bands of ultraviolet to those found to be therapeutically effective.[14]

3. Levine reports that plain white petrolatum similarly narrows the ultraviolet bands. The petrolatum is massaged into the psoriatic lesions and surround skin areas. Exposure is then administered at scheduled dosages[15] (Table 3-4).

Traditional Technique for Psoriasis

1. After their individual MED is established, most psoriasis patients are seen two to three times weekly.

2. If strong erythema develops at this dosage or frequency, diminish intensities by distance or angulation, shorten duration, or reduce the frequency of treatment.

3. Patients often report an immediate decrease in itching following the first or second exposure. Do not be discouraged if this does not occur, since psoriasis is a condition that is very difficult to manage and one that usually is a long-term problem for both the patient and the physical therapist.

4. In many instances, patients eventually purchase a lamp to be used at home with guidance from the physical therapist.

The PUVA Technique[16]

PUVA, in contrast, is a strictly controlled technique and is rarely, if ever, recommended for use at home. Patients are required to ingest a photosensitizing medication (usually psoralen) 2 hours prior to exposure to ultraviolet radiation. Patients should not be exposed to sunlight for approximately 8 hours following the oral medication.

Clinical exposure to ultraviolet is shortened to 10 to 15 seconds, with very few exposures ever exceeding the 1-minute mark. The ultraviolet source must emit radiation in the A band, or 360 nm; therefore, the lamp must be specially designed and purchased with this technique in mind.

Total Body Technique

When total body exposure is indicated, "quartering" of the patient is recommended to ensure equal exposure throughout.[17]

1. Have the patient disrobe and lie prone; then drape the patient's body from the waist down, exposing the upper torso for the scheduled dosage.

2. Move the drape sheet to cover the upper portion of the patient's body, leaving the lower portion to be irradiated.

3. Have the patient turn to the supine position, with the upper torso draped from the waist for scheduled exposure.

4. Last, move the drape to cover the lower anterior body, exposing the upper chest and face, taking care as suggested with the eyes, nose, nipples, and abdominal high points. These areas may be covered with moist gauze squares or additional petrolatum if necessary.

With the above technique, approximately equal exposures may be prescribed for all four quarters.

TABLE 3-4. Treatment Frequency

Dosage	Regimen	Effect
SED (suberythemal dose)	Daily	No erythema
MED	Daily or every other day	Slightly pink
1 ED (first degree)	2–3 days	Red, itch, "sunburn"
2 ED (second degree)	10–14 days	Same (exaggerated) as above
3 ED (third degree)[a]	Local treatment *only*	Edema, blisters

[a]Used in special cases only with special lamps for this purpose.

FIG. 3-9. Spot quartz (Kromeyer-type) lamp in the 253.7-nm band, producing a highly bactericidal effect with little pigmentation. A Wood's filter for "black light" fluorescent application is shown. (Courtesy of Birtcher Corp., El Monte, California.)

SPECIAL LAMPS

Kromeyer Lamp

The Kromeyer lamp (Fig. 3-9), a special type of ultraviolet device, emits radiation in the far bands and is utilized in the management of open lesions, ulcerations, and internal targets. It is used only for local treatment, never for total body techniques. The radiating area is rather small compared with that of standard ultraviolet lamps, sometimes only 3 inches (7.5 cm) in diameter. For orificial treatments (i.e., mouth, open wounds) special equipment is added to narrow the diameter of the beam to about an inch. The lens is pressed against the skin to create a localized, superficial ischemia, allowing increased penetration of the radiation. The destruction of target tissue is a delicate and important function of this type of ultraviolet apparatus; it should be utilized only as part of an overall dermatologic and surgical management approach.

Wood's Light

The Wood's light is a lamp that produces radiation in the ultraviolet band of 350 to 400 nm and is used as a diagnostic tool rather than a therapeutic device. It identifies fluorescent microorganisms, facilitating differential diagnoses.

Phototherapy

Ultraviolet radiation has proven valuable in the nursery for the prevention of postnatal jaundice.[18,19] Although ordinary fluorescent bulbs have been utilized for this purpose, specialized sources are available with specific wavelengths and intensity parameters.

COLD LASER

The use of light for therapeutic purposes dates back to the ancient Greeks, Romans, and Egyptians. Current research into the physiologic benefits of light therapy has developed an area of great interest: the laser. Most research in the uses of lasers is reported by European sources.[20–23] Only during the past decade have American researchers begun to add the results of their studies.[24,25]

The popularly used term *laser* is an acronym for light amplification by stimulated emission of radiation. The power levels of light are greatly amplified by the emissions of radiation from stimulation of specific substances. Every substance radiates emissions in varying wavelengths and frequencies. Helium, neon, cobalt, and carbon dioxide are examples of substances that, when irradiated, have application in commercial, medical, and engineering processes. In therapeutic use of laser, we are concerned with the emissions of a helium-neon mixture in the 632.8-nm band of the spectrum. Other medically useful lasers are in the 904 nm (infrared) band: GaAs, 854 nm; Nd_3 + YAG (yttrium-aluminum-garnet), 106 nm; Ar, 488 nm; CO_2, 1060 nm; and the ruby, 694.3 nm. Most of these, however, are used in surgical procedures. The HeNe 632.8-nm laser, however, is utilized by physical therapists, as explained below.

MECHANICS

To obtain the laser, a tube filled with a gaseous mixture of helium and neon is stimulated electrically to emission levels. Within the highly reflective, polished walls of the tube, the molecules reverberate and carom off the walls in a highly agitated state, building energy as they do so. When a critical level is reached, the flow of energy literally "bursts" through the semisilvered (similar to a one-way mirror) front end of the tube and is channeled along an optic fiber to the beam applicator, or probe, for clinical applications (Fig. 3-10).

PHYSICS

Intensity

One might naturally assume that the laser generates tremendous power; however, the type of laser used by physical therapists has the intensity of only 1 mW—less than the power of a 60-W light bulb held a few inches from the face! Commercial and industrial lasers (i.e., hot lasers) range in the thousands and millions of watts and are used for cutting, drilling, and destructive applications. The tool used by physical therapists, a cold laser, is used primarily for healing and other nondestructive purposes.

Physical Characteristics

Paraphrasing the traditional Passover ceremonial question, we may ask: "Why is this light different from all other lights?" There are three characteristics of laser light that clearly differentiate it from ordinary light.[26]

Monochromaticity

Ordinary light comprises a conglomeration of many wavelengths, commonly known as ROYGBIV, or the visible spectrum of red, orange, yellow, green, blue, indigo, and violet, all merging to produce "white" light. Laser light, however, consists of one wavelength only.[27] In the case of our therapeutic unit, the band is 632.8 nm. Because this wavelength falls within the R section of the visible spectrum (390 to 770 nm), the laser light of 632.8 nm is a brilliant red color (Fig. 3-11).

Coherence

Because the wavelengths of ordinary light are so variable and do not "match" in waveforms,

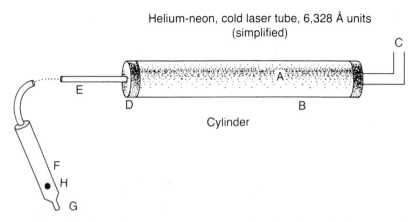

Helium-neon, cold laser tube, 6,328 Å units
(simplified)

FIG. 3-10. Laser mechanics: When the HeNe mixture **(A)** within the silvered tube **(B)** reaches maximum excitation levels from external electrical stimulation **(C)**, the laser radiation emerges at the semisilvered end of the tube **(D)**, passing along an optic fiber **(E)** into the probe housing **(F)** to the probe tip **(G)**, where a trigger **(H)** releases the beam for application.

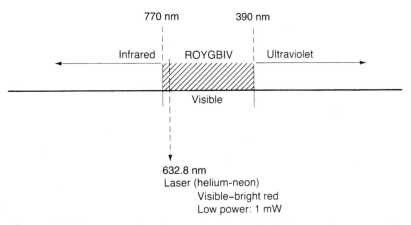

FIG. 3-11. Monochromaticity: lasers contain only one wavelength. ROYGBIV—red, orange, yellow, green, blue, indigo, violet.

frequencies, or shapes, there is much scrambling of waveforms, cancellations and reinforcement of individual waves, and interference in the production of energy in general; this factor minimizes the power of ordinary light as an energy source. The identical wavelengths and forms that comprise laser light cause it to be greatly amplified, since the "waves and troughs" of the radiation are reinforced. Because they are parallel and in line with each other, they are termed *coherent* (Fig. 3-12).

Nondivergence

The laser beam is unique in the absolute "straightness" of the directed radiation. Ordinary light shines in all directions (e.g., consider a light bulb radiating in all directions). The sun is another example of omnidirectional radiation. The laser, on the other hand, shines in only one direction, not unlike a flashlight, although its beam is far more concentrated and narrowed. The divergence of a laser beamed to the surface of the moon from Earth showed a deflection of just a

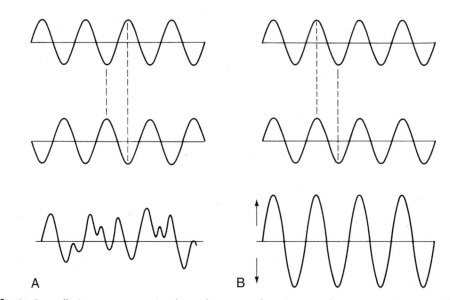

FIG. 3-12. (A) Cancellation: waves not in phase do not produce increased power or concentrated frequency responses. **(B)** Coherence: laser waves are in phase, increasing power by reinforcement.

few meters after a journey of more than 260,000 miles (416,000 km) (Fig. 3-13).

PHYSIOLOGY

Wound Healing

Ordinary light does not penetrate the skin or underlying tissues. Only the retina can absorb the visible spectrum. As discussed previously, infrared is absorbed at the 3-mm level and ultraviolet at the 1-mm level. Research has indicated that the 632.8 nm wavelength of the cold laser may stimulate intracellular structures and functions. In the human cell, mitochondria are responsible for the production of DNA, RNA, and other metabolism-related substances. What better place to direct our efforts than to the source of building materials. One of the prime applications of laser light in physical therapy is that of wound healing.[28-30]

Analgesic Effects

In pain control, the penetration of laser light is often compared with an acupuncture needle. Therefore, when the laser is used for analgesic purposes, the beam is generally directed at acupuncture points, trigger points, and nerve roots, as is done with acupuncture. (With acupuncture-like techniques, it is the target point that is of great importance. Stimulation for analgesia or other physiologic effects may be heat [moxabustion], soundwaves, digital pressure [shiatsu], electricity [transcutaneous electrical nerve stimulation—TENS], and, with the laser light).[31,32]

Penetration

Direct penetration of the 632.8-nm HeNe cold laser at 1 mW is said to be approximately 0.8 mm; indirect penetration after refraction, dispersion, reflection, and partial absorption is 10 to 12 mm. As previously noted, laser light behaves according to all the known laws of optics listed previously, as well as according to the inverse square and cosine laws with regard to intensity levels (Figs. 3-14 and 3-15).

Absorption

Absorption apparently depends on the resonance of the tissues or on the subject's water content. Many of the penetration characteristics and limitation factors are determined by fluids (e.g., water, blood).

PHYSICAL EFFECTS

Heating

A mild but reversible heat is produced with the cold laser. The tissues revert back to the pre-lasing temperatures immediately following radiation. Although the thermodynamics involved do not produce a therapeutic level, cell-wall permeability has been shown to be a favorable result of heat and may play a role in the reaction of the laser on the cell wall.

Dehydration

Loss of water following radiation is another reversible process. Such loss may be attributed to

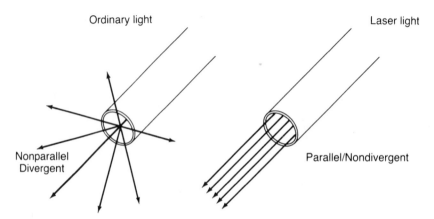

Ordinary light   Laser light

Nonparallel Divergent   Parallel/Nondivergent

FIG. 3-13. Nondivergence: laser radiation is extremely straight; all rays are parallel.

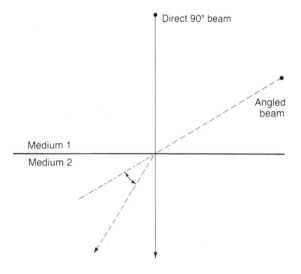

FIG. 3-14. Refraction: bending of the radiant waves as they pass from one medium to another.

the minor heating and/or the transfer of fluids to distant sites. Apparently, it does not play a major role in the laser's effectiveness.

Coagulation of Proteins, Thermolysis, and Evaporation

Coagulation of proteins, thermolysis, and evaporation are irreversible and should *not* occur with the dosages and techniques used by physical therapists. *Coagulation* is a permanent process, comparable to that of an egg frying in a skillet. *Thermolysis,* or "melting from heat," is also a lasting state. *Evaporation,* of course, involves the transformation of liquids to the gaseous state and is not easily reversed. With competent clinicians and sophisticated equipment, none of these conditions should be produced in routine administration of the cold laser. With higher power levels,

one must be cautious and aware of the above factors.

DOSAGES

Actual dosages with the cold laser depend on the power factor, duration of radiation, and tissue resonance. Focusing the beam properly will bring into play the inverse square and cosine laws, so that distance-to-target and angulation of beam-to-target will affect the dosage administered.

Wound Healing

In most wound healing applications, the prescribed dosage is 90 s/cm^2 of open lesion. This may require several minutes of hand-held direction of the beam over the surface of the wound so

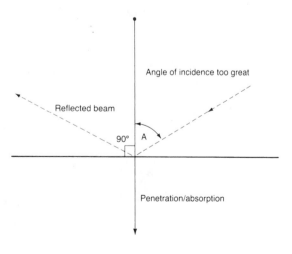

FIG. 3-15. Reflection: when the angle of incidence is too shallow, the radiation will be reflected or "bounced back," usually at the same angle.

that each square centimeter is exposed for the same 90 seconds. The probe tip is held approximately 2 to 3 mm from the surface to obtain a "disc" of laser light about 1 cm in diameter on the surface of the wound. The nondivergent nature of the laser is modified by a lenslike spreading of the beam for therapeutic purposes.

Pain Control

At acupuncture trigger points, nerve roots, and pain sites, a dosage of 15 to 30 seconds for each point is recommended. Unlike in the open wound technique, the probe tip is held in contact with the skin at these points during the procedure.

MODES

Current laser equipment offers the clinician two modes: *continuous* and *pulsed* beams.

Continuous Beam

The continuous mode is recommended for acute pain and fresh wounds.

Pulsed Beam

The pulsed mode has been found more effective with chronic conditions. Pulsed models vary from 1 to 80 pulses per second, depending on the manufacturer. One suggested technique for chronic pain or long-standing open lesions would be in the range of 4 to 10 pulses per second.

PARAMETER SETTINGS

Current laser units offer manual timing or automatically timed treatments. For longer exposures, such as with large open wounds, use of the automatic timer is suggested. For pain control, either manual or automatic timing may be utilized. Power levels are preset by the manufacturer and the FDA at 1 mW. The power is reduced to half (0.5 mW) when the pulsed mode is in effect. Later models will offer high power availability in the 5 mW range. Infrared lasers in the 904 nm band are also available.

PROBE DATA

The cold laser equipment includes a probe, from which the laser beam is emitted through an

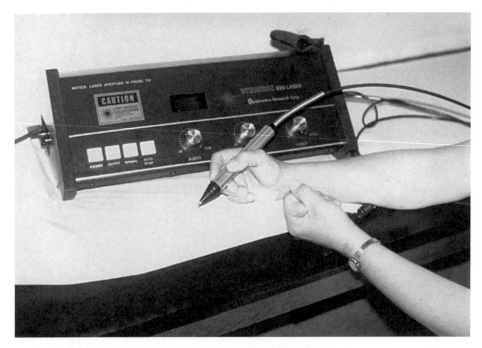

FIG. 3-16. Clinical laser used by the author. Note probe and ground cylinder. (Dynatron 820, HeNe 632.8 nm.) (Courtesy of Dynatronics Laser Corp., Salt Lake City, Utah.)

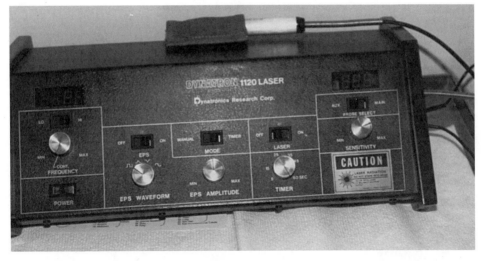

FIG. 3-17. A recent addition to the cold laser is the availability of microamperage electrical stimulation, either simultaneously with the laser radiation, or separately, if desired. Note the selection of square, spike, or sinusoidal waveforms on the control panel. (Courtesy of Dynatronics Laser Corp., Salt Lake City, Utah.)

opening in the tip (Fig. 3-16). This tip also serves as a differential resistance device (ohmmeter) to locate the points of lowest resistance for acupuncturelike applications of the laser. (Acupuncture points are points of low electrical resistance compared with the surrounding skin.) A visual, digital read-out on a lighted electronic display (LED) is available to the clinician, along with an audible signal, preset to sound at a specific microampere level (e.g., 30 mA on the LED). This is extremely helpful when the clinician is searching with the LED for target points that are out of the range of sight.

The probe, when utilized for point searching, requires the use of a reference electrode, usually a hand-held, metallic cylinder, connected to the chassis by a separate but integrated circuit. When this cylinder is used, the patient grasps the ground cylinder while the physical therapist searches the patient's skin for proper points and then applies the laser beam to them. The reference is not necessary when treating open wounds, since there is no skin contact during the treatment and the target site is obvious. The audible volume of the signal is controlled by the clinician.

Some of the more recently marketed units also include simultaneous or separate electrical stimulation modes with the laser circuitry (Fig. 3-17). The tip on these units functions as an electrode, and the reference cylinder functions as a secondary electrode to complete the circuit. Various waveforms of electrical stimulation (e.g., square

wave, sine wave, spike wave) are offered on these units.[27]

When the 904-nm infrared laser is available as an optional wavelength with units, the probe may also contain a guide light for operation, since the infrared beam is invisible.

INDICATIONS

Cold laser is indicated[33-38] for treating

1. Open lesions (Fig. 3-18)
2. Decubitus ulcers (see Fig. 3-18)
3. Diabetic ulcers (Fig. 3-19)
4. Lacerations (Fig. 3-20)
5. Incisions (Fig. 3-21)
6. Burns (Fig. 3-22)
7. Chronic and acute pain (Figs. 3-23 and 3-24)
8. Restricted joint ranges of motion (Fig. 3-25)

CONTRAINDICATIONS

1. Do not radiate the eye directly.
2. Whether pregnancy is a contraindication to use of the laser has not yet been determined, but the laser is suspect because of its mobilizing effect on steroids in the human system.
3. Do not use the laser with patients who are naturally photosensitive or who are photosensitized by medications.

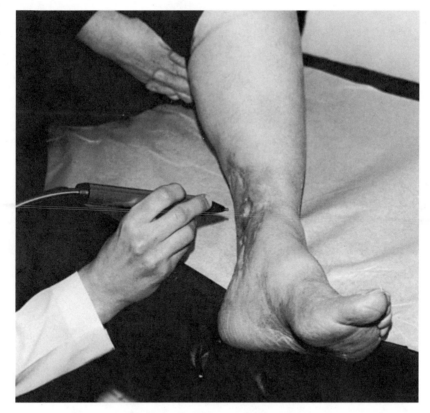

FIG. 3-18. Cold laser application to extensive open lesion (cellulitis). Actual treating distance is about 1 to 2 mm from the surface.

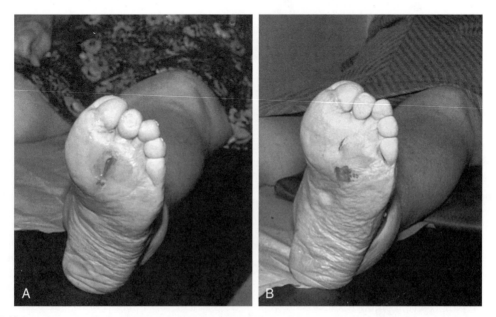

FIG. 3-19. A, Open lesion, diabetic, prior to lasings (August 23, 1983). **B,** Lesion following laser radiation, twice weekly (October 19, 1983). (From Kahn,[29] with permission.)

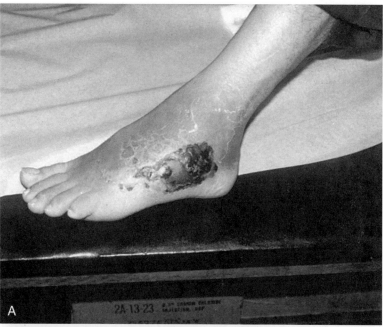

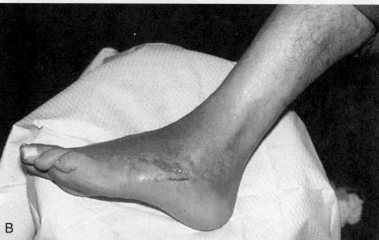

FIG. 3-20. A, Traumatic lesion, with thick eschar, prior to lasing (September 3, 1982). **B,** Same lesion following lasings, twice weekly (November 24, 1982).

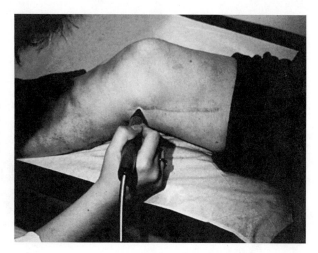

FIG. 3-21. Cold laser radiation as applied to a postsurgical scar. Sensitivity and discoloration diminish with continued treatment. Thirty seconds per linear centimeter is a recommended dosage in this instance.

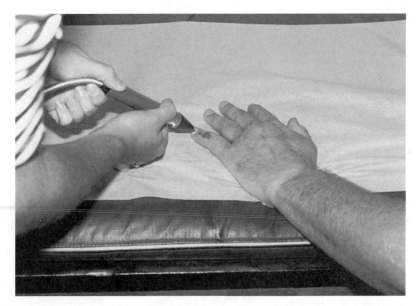

FIG. 3-22. Laser applied to slow-healing burn. Note two-handed grip for steady administration.

PRECAUTIONS

Poor results may ensue in those patients

1. Of extreme age
2. Under heavy medication
3. With thick eschar
4. With considerable scar tissue
5. With extremely dry skin
6. With active infection

A touch of moisture on the tip of the probe or the target skin may enhance the electrical contact needed for efficient point searching. Perspiration or other skin moisture will naturally give false read-outs; adjustments must be made in such circumstances. Dry the patient's skin prior to lasing and/or select higher readings on the LED for targeting.

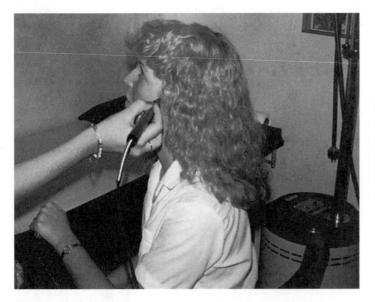

FIG. 3-23. Laser directed to a painful temporomandibular joint (TMJ). Note ground cylinder gripped by patient, here used to locate appropriate points accurately, since probe tip is in contact with patient's skin for lasing.

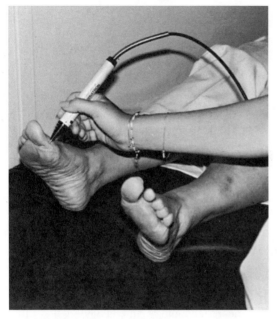

FIG. 3-24. Cold laser applied directly to skin over a painful hallux and associated points.

TREATMENT PROCEDURES

Tissue Healing

Ordinary Wounds

1. No ground is required.
2. Select continuous mode if the wound is fresh; use pulsed mode at 10 pulses per second if the wound is old.
3. Radiate for 90 s/cm^2 at a distance of 2 mm from the patient's skin, with slow movement of the probe to cover the entire wound area. The red disc that appears on the skin should be about 1 cm in diameter (see Fig. 3-21).

Extensive Wounds

1. Radiate the perimeter of the wound with a slowly moving probe.
2. Cover each centimeter along the perimeter for 90 seconds.
3. As the wound closes (heals) with subsequent lasings, the central portion may be treated as described above (see Fig. 3-18).

Grid Technique

1. The grid technique is also recommended for the extensive lesions requiring long exposures.
2. Mentally "grid" the wound into 1-cm^2 sections and proceed to apply the laser to each section for the suggested 90 seconds (Fig. 3-26).

Frequency of Treatment

With open wound procedures, treatment is administered two or three times weekly; results will be visible to the casual examiner within the first three or four treatments. Patience is required with larger wounds, scarred areas, eschar, tough skin, and infections.

Pain Control

1. Use probe and ground cylinder.
2. Search the patient's skin in the area of known acupuncture points, palpable trigger points, nerve roots, and pain sites by lightly touching the probe tip to the skin (while the patient grips the ground cylinder) (Fig. 3-27).
3. Monitor the LED for high read-outs, usually greater than 20 µA, or listen for the audible signal at approximately 30 µA on the LED screen.
4. When the audible signal is heard, press the trigger on the probe handle to activate the laser.
5. Hold in contact with the point for 15 to 30 seconds.
6. If the automatic mode is in effect, it is not necessary to hold down the trigger, since the laser will be "on" for the selected number of seconds on the timer dial.
7. Small, reddish marks will appear at the points of probe contact, especially if the physical therapist has a "heavy hand"; however, the marks should disappear within 1 hour and cause no reason for concern.

Frequency of Treatment

1. Administer treatments daily if necessary as indicated by the severity of pain.
2. If favorable results are not noted by the third or fourth treatment, consider other modalities for analgesia.

RECENT DEVELOPMENTS

The latest developments in the cold laser field have resulted in models that combine microampere electrical stimulation at the probe tip with

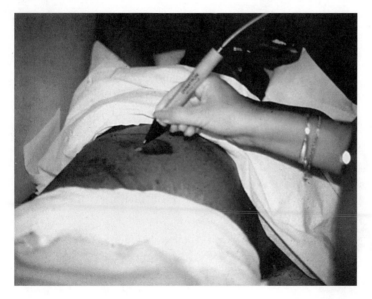

FIG. 3-25. Cold laser applied to long-term traumatic torso keloids in an effort to reduce the sensitivity and severe pruritus. Recommended dosage in this instance was 90 s/cm^2.

the laser radiation for enhanced analgesic effects. In addition, models featuring higher milliwattage power (5 mW) levels are also being offered. Cold laser equipment is now prevalent in Canada to a much greater extent than in the United States, and offers clinicians considerably higher power levels (i.e., 10 to 20 mW).

The application of cold laser radiation to points in the ear—*auricular therapy*—is rapidly becoming more popular among innovative clinicians (Fig. 3-28). Both as an evaluative tool and a therapeutic technique, auricular therapy has been found effective in determining and reducing pain in distant anatomic locations (as with acupuncture [which also utilizes ear points]). Points on the ear correspond to anatomic locations analogous to the fetal-like configuration of the outer ear (Fig. 3-29). Originating in France after WWII,[39] this procedure has slowly been incorporated into American technology and is undergoing close investigation by clinicians across the country.[40]

Unfortunately, the status of the HeNe cold laser is still on hold. Premarket approval from the FDA has not been forthcoming despite clinicians' efforts and those of the manufacturers. This modality is used throughout the world, and its success is well documented. In the United States, however, it can only be used at this time under clinical investigational status. I recently lectured in Buenos Aires, Argentina, where I found lasers to be used routinely by clinicians. The cold laser is a commonly used modality in Canada and is available in varying intensities, often much greater than those available in the United States.

One of the better reference books on lasers is from England: G. David Baxter's *Therapeutic Lasers* certainly covers the field well and makes

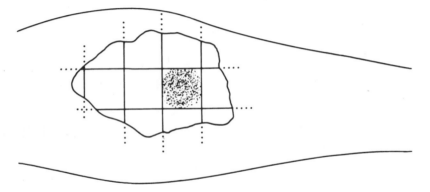

FIG. 3-26. Grid technique.

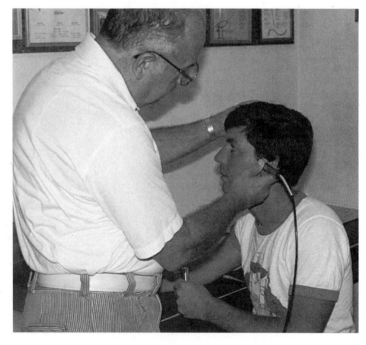

FIG. 3-27. Similar to the temporo-mandibular joint technique, this application of the laser is for a trigeminal nerve pain *(tic douloureux).* Note reference cylinder gripped by patient for accurate point location.

an excellent case for acceptance.[41] The FDA requires many documented studies of success for a *single* diagnosis, rather than the reams of supportive statistics that have been submitted covering a wide range of conditions. The last I have heard is that the condition selected was "rheumatoid arthritis." It is hard to believe that this condition was chosen when significant results with the cold laser have been provided on enhancement of open wounds, a much more common use for the laser. Because I see so few patients with rheumatoid arthritis in my limited retirement practice, I've been unable to contribute documented studies to the FDA. However, it is still possible to use the cold laser under "clinical investigative" status, which makes reimbursement almost impossible from most insurance carriers.

I am still hoping that the FDA will approve the cold laser for general use in the near future. It

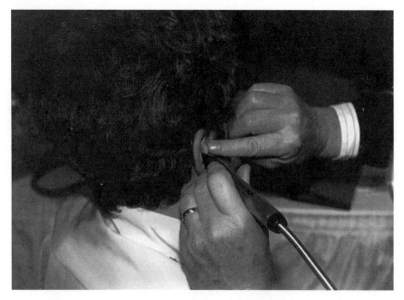

FIG. 3-28. Cold laser administered to specific ear points. (Courtesy of Dynatronics Laser Corp., Salt Lake City, Utah.)

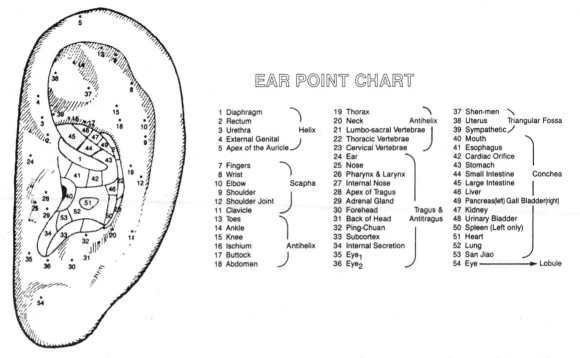

EAR POINT CHART

1 Diaphragm	19 Thorax	37 Shen-men
2 Rectum	20 Neck	38 Uterus
3 Urethra	21 Lumbo-sacral Vertebrae	39 Sympathetic
4 External Genital	22 Thoracic Vertebrae	40 Mouth
5 Apex of the Auricle	23 Cervical Vertebrae	41 Esophagus
	24 Ear	42 Cardiac Orifice
7 Fingers	25 Nose	43 Stomach
8 Wrist	26 Pharynx & Larynx	44 Small Intestine
10 Elbow	27 Internal Nose	45 Large Intestine
9 Shoulder	28 Apex of Tragus	46 Liver
12 Shoulder Joint	29 Adrenal Gland	49 Pancreas(left) Gall Bladder(right)
11 Clavicle	30 Forehead	47 Kidney
13 Toes	31 Back of Head	48 Urinary Bladder
14 Ankle	32 Ping-Chuan	50 Spleen (Left only)
15 Knee	33 Subcortex	51 Heart
16 Ischium	34 Internal Secretion	52 Lung
17 Buttock	35 Eye$_1$	53 San Jiao
18 Abdomen	36 Eye$_2$	54 Eye ⟶ Lobule

Antihelix — 19–23; Helix — 1–5; Scapha — 7–12; Antihelix — 13–18; Triangular Fossa — 37–39; Tragus & Antitragus — 28–31; Conchea — 40–53

FIG. 3-29. Note the resemblance of the ear configuration to the human fetus. (Courtesy of Techmed Corp., Eden Prairie, Minnesota.)

would be a tremendous asset to clinical practice and of great benefit for the patients. Perhaps this manual, now available in Spanish and Polish, will stimulate more clinical interest in the cold laser and provide the needed data for eventual approval for this valuable addition to the clinical armamentarium in the United States (see Fig. 3-30).

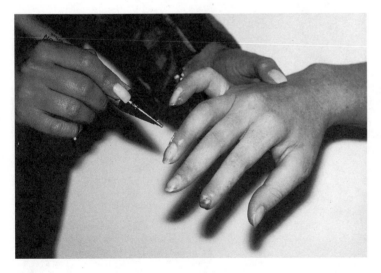

FIG. 3-30. The HeNe cold laser administered to a patient with scleroderma in an effort to prevent loss of the terminal phalanx (as noted in the index finger). (Courtesy of Dynatronics Laser Corp., Salt Lake City, Utah.)

REFERENCES

1. Shriber WJ: A Manual of Electrotherapy. 4th Ed. Lea & Febiger, Philadelphia, 1981, p. 15
2. Wadsworth H, Chanmugam A: Electrophysical Agents in Physiotherapy. Science Press, Marrickville, Australia, 1980
3. Forster A, Palastanga N: Clayton's Electrotherapy. 9th Ed. Balliere Tindall, London, 1985
4. Wilensky J, Aronoff GM: Therapeutic Use of Heat. R$_x$: Home Care, July 1985, p. 70
5. Artuson E: Infrared helps to control burn hypermetabolism. Reprinted in Medical Tribune, May 5, 1976
6. Montgomery PC: The compounding effects of infrared and ultraviolet irradiation upon normal human skin. Phys Ther 53:5, 1973
7. Licht S: Therapeutic Electricity and Ultraviolet Radiation. E. Licht, New Haven, CT, 1959
8. Kellogg JM: The electric light bath. JAMA, December 24, 1898; reprinted in JAMA 226:1533, 1973
9. Mitchell AJ: UV and Alopecia. Nat Clin Derm Conf, Chicago, September 15, 1983
10. Bierman WS, Licht S: Physical Medicine in General Practice. 3rd Ed. Hoeber, New York, 1952
11. Anonymous: Ultraviolet: Syllabus. 3rd Ed. Burdick Corp., Milton, WI, 1976
12. Merchant M, Hammond R: UV for pityriasis rosea. Cutis 14:548, 1974
13. Ramsay CA: UV and blood flow. Br J Dermatol 94:487, 1976
14. Perry HP, Soderstein LW, Schulze RW: Goeckermann regimen for psoriasis. Arch Dermatol 98:178, 1968
15. Levine MJ: Use of petrolatum with UV for psoriasis. Am Acad Dermatol (San Francisco), reprinted in Medical World News, January 8, 1979
16. Steck WD: PUVA in psoriasis. Consultant, April 1977, p. 130
17. Gilchrist BA: The human sunburn reaction—an update. Med Times 112(6):92, 1984
18. Eggert LD: Phototherapy. West Soc Ped Res, Carmel, CA, March 1984; reprinted in Medical World News, March 1984, 25
19. Brown AK, Lucey JF: Phototherapy. Am Acad Pediatr, San Francisco, 1979; reprinted in Medical World News, November 26, 1979
20. Wolbarscht ML (ed): Laser Applications in Medicine and Biology. Plenum Press, New York, 1977
21. Bogatryov V: The cold laser. Soviet Life, No. 3 (258), March 1978
22. Bischko JJ: Acupuncture and electrotherapeutics. Res Int J 29:1980
23. Caspars KH: Stimulation therapy with laser beams. Physikal Med Rehabil 18(9):426, 1977
24. Goldman JA: Investigative studies of laser technology on rheumatology and immunology. Lasers Med Surg 1:93, 1980
25. Parrish JA: Photomedicine—potential for lasers; an overview. Proc Am Soc Laser Med Surg, January 1982
26. Kleinkort J, Foley RA: The cold laser. Clin Management 2:30, 1982
27. Anonymous: "The light emitted by this type laser (HeNe 6328 AU) is in the red field of the color spectrum. This falls within the spectral range of highest cell transparency." Quoted from specifications of the 1120 Laser unit, Dynatronics Corp., Salt Lake City, UT, 1985
28. Mester E: The effect of laser rays on wound healing. Am J Surg 122:532, 1971
29. Kahn J: Case reports: open wound management with the HeNe (6328 AU) cold laser. JOSPT 6:203, 1984
30. Shaw CJ: The Effects of Low Power Lasers on Wound Healing: a Review of the Literature. New York University, PT program student paper, December 1982
31. Snyder L, Meckler C, Borc L: The effect of cold laser on musculoskeletal trigger points: a double blind study, abstracted. Phys Ther, May 1984, p. 745
32. Barnes JF: Electro-acupuncture and cold laser therapy as adjuncts to pain treatment. J Craniomandibular Pract 2:148, 1984
33. Michlovitz S: Thermal Agents in Rehabilitation. FA Davis, Philadelphia, 1987
34. Snyder-Mackler L, Barry AJ, Perkins AI, Soucek MD: Effects of helium-neon laser irradiation on skin resistance and pain in patients with trigger points in the neck or back. Phys Ther 69:336, 1989
35. Gogia PP, Hurt BS, Zirn TT: Wound management with whirlpool and infra red cold laser treatment. Phys Ther 68:1239, 1988
36. Snyder-Mackler L, Bork CE: Effect of helium-neon irradiation on peripheral sensory nerve latency. Phys Ther 68:223, 1988
37. Anonymous: Healing with Light—Medical Forum. Harvard Medical School Newsletter, February 1988, p. 3
38. Kahn J: Cold Laser (videotape). AREN, Pittsburg, PA, May 1988
39. Nogier P: Treatise on Auriculotherapy. Lyon, France, 1973
40. Oleson TD: UCLA Research on Auriculotherapy. 2nd Ed. ARC, Los Angeles, 1987
41. Baxter GD: Therapeutic Lasers. Churchill Livingston, London, 1994

4

Ultrasound

STANDARD ULTRASOUND

Unlike the electrical modalities, ultrasound is unique in that the longitudinal waveform associated with sound is not electromagnetic in nature.[1] It is similar to the child's toy Slinky in form (accordianlike action). Soundwaves represent the compression and refraction of a vibrating medium (Fig. 4-1). Electromagnetic waveforms can be transmitted across a vacuum, such as interplanetary spaces, but soundwaves require a medium for transmission. As a waveform, sound follows the rules of physics regarding reflection, absorption, refraction, and dispersion (see Figs. 3-14 and 3-15). At times, this longitudinal waveform becomes transverse and may present problems of unwanted heat buildup. This is discussed later in the chapter.

Ultrasound has been found useful as a therapeutic modality for the reduction of muscular and tendinous spasm.[2] It has also been utilized for pain and other pathologic conditions through the ability of soundwaves to introduce molecules of chemical substances through the skin by a process called *phonophoresis*.[3] Because water is an excellent conduction medium, subaqueous techniques with ultrasound permit sonation of anatomic regions difficult to reach with standard techniques (e.g., fingers, toes, and bony prominences of the elbow, ankle, and wrist). Both phonophoresis and underwater ultrasound are discussed later in this chapter.

PHYSICS

In the United States, therapeutic ultrasound is limited to the frequency of approximately 1 MHz.[1] This frequency is reached by the transformation of household current of 60 Hz/110 V alternating current to 500 V or more by electronic components in the ultrasound apparatus. The higher voltage is then applied to oscillators or vibrators that boost the household frequencies to the desired level of 1 MHz.

Piezoelectricity

The higher frequency is then imposed on a crystal with piezoelectrical qualities.[4] Piezoelectricity is a natural phenomenon found in certain mineral crystals, such as germanium and quartz, but it may also be synthesized commercially, for example, lead-zirconium-titanate (PZT). This crystal transforms mechanical energy into electrical energy and its reverse, electrical into mechanical (Fig. 4-2). If a piezoelectric crystal, natural or synthetic, were to be compressed or deformed by mechanical means, a small electrical charge would result within the crystal; conversely, if an electrical charge were to be imposed on the crystal, a vibration of mechanical deformity of the molecular structure of the crystal would ensue. (Crystals differ from other amorphous substances due to the unique configuration of their molecular structure and the mathematically precise arrangement of their component atoms. Changes in the subatomic forces by deformation lead to electrical discharges due to disruption of normally balanced intramolecular forces.) Human bone exhibits piezoelectrical characteristics, a fact put to application in the healing of nonunited fractures with electronic assistance (see Ch. 6).[5,6]

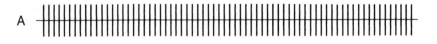

FIG. 4-1. The longitudinal waveform of ultrasound (A) compared with the transverse waveform of electromagnetic radiation (B).

Transformation of Electrical Energy into Sound

The high-frequency current of 1-MHz alternating current is imposed on the PZT crystal of the transducer or soundhead, thereby transforming the energy in the crystal to a vibration—or deforming oscillation of the molecular structure of the crystal at the same frequency, 1 MHz. The crystal is held by cement in contact with the metallic or plastic faceplate of the transducer and consequently causes vibration of the housing at the same 1-MHz frequency. Thus, the electrical energy of the original current is transformed into a vibratory soundwave—ultrasound—since it is beyond the normal range of audible soundwaves, which are at approximately 25 to 15,000 counts per second.

Conduction of the Wave Energy

Substances in contact with the faceplate, such as water, oils, and gels, conduct the wave energy to adjoining surfaces, including human skin. This process is repeated from tissue to tissue as long as there is a conducting medium between them. Because air is a very poor sound conductor, it must be replaced by a more effective medium when used therapeutically.

Penetration and Absorption

Ultrasound waves have been reported to penetrate as deep as 4 to 6 cm into the tissues.[7] The factors of absorption, refraction, reflection, and dispersion must be considered at all times, with longitudinal or transverse wave propagation, however. Tissues with high fluid content, such as blood and muscle, will absorb soundwaves better than will less hydrated tissues. Although seemingly not hydrated, nerve tissue is an excellent conductor of ultrasound waves.[8] Under Techniques in this chapter, the suggested sounding of nerve roots in association with peripheral conditions is discussed.

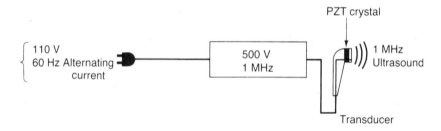

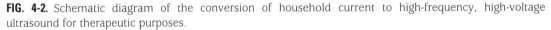

FIG. 4-2. Schematic diagram of the conversion of household current to high-frequency, high-voltage ultrasound for therapeutic purposes.

PHYSIOLOGIC EFFECTS

Ultrasound has four basic physiologic effects.[4]

Chemical Reactions

Just as a test tube is shaken in the laboratory to enhance chemical reactions, ultrasound vibrations stimulate tissue to enhance chemical reactions and processes therein and ensure circulation of necessary elements and radicals for recombination.

Biologic Responses

The permeability of membranes is increased by ultrasound, which enhances transfer of fluids and nutrients to tissues. This quality is of importance in the process of phonophoresis, wherein molecules are literally "pushed" through the skin by the soundwave front for therapeutic purposes.

Mechanical Responses
Cavitation

The high-frequency vibration of ultrasound deforms the molecular structure of loosely bonded substances. This phenomenon is therapeutically useful for the sclerolytic effects produced in the attempt to reduce spasm, increase ranges of motion due to adherent tissues, and break up calcific depositions, mobilizing adhesions, scar tissue, and so on. If used to extremes of power or duration, this deforming mechanism can collapse the molecules and cause destruction of the substances—a phenomenon called *cavitation*.[9] This should not occur with the power levels and treatment parameters used clinically by physical therapists. The presence of unwanted collagenous material, in the conditions just described, makes this semidestructive force valuable, however, since these substances are held loosely in molecular bondage and can be disrupted by the ultrasound energy.

Tendon Extensibility

The sclerolytic action of ultrasound apparently increases the extensibility of tendons, crucial to those that have been shortened by inflammation, strain, or disease, and provide the clinician with an excellent formula for the management of spasm.

Demonstrating Mechanical Effects

Two simple demonstrations will indicate the mechanical effects of ultrasound without thermal increases.

Experiment 1

Cup the palm of one hand so that a small amount of lightweight mineral oil can be held there. Apply ultrasound transmission gel to the head of the transducer, and sound the dorsum of the upturned palm. You will notice movement on the surface of the oil in the palm. The ultrasound waves have penetrated through your hand to affect the oil with no appreciable rise in temperature of the skin or lesions of your hand.

Experiment 2

Pour several drops of water onto the upturned face of a transducer in operational status and note the bubbling of the water without an obvious increase in the temperature of the water. Neither is there any noticeable rise in temperature when ultrasound is applied subaqueously. Ripples are, however, seen on the surface of the water, indicating the mechanical manifestation of the ultrasound energy (Fig. 4-3).

Thermal Effects
Heating Benefits

Many academicians and clinicians list the thermal qualities of ultrasound as the most important of the four physiologic effects, often referring to the treatment as *ultrasonic beat* or *ultrasonic diathermy,* the latter being a gross misnomer.[10] The fact that rapid oscillations of molecules will cause a buildup of heat is accepted by all trained physical scientists.[11] Those of us who consider the mechanical effects of ultrasound to be the prime factor in clinical success believe that the highly selective absorption behavior of ultrasound limits its efficacy as a therapeutic heating agent, however.[12–16] (The skin obviously becomes warm when massaged, but such massage can hardly be called a heat treatment.)

It must be remembered that heat buildup with ultrasound is localized to the tissues directly under the transducer head. Since the accepted technique is to keep the soundhead moving

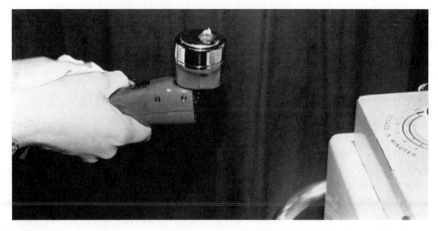

FIG. 4-3. Water bubbles when dripped onto the soundhead, without noticeable change in the temperature of the water, indicating the mechanical effects of ultrasound independent of thermal increases.

during treatment, it is highly unlikely that any appreciable therapeutic heating will take place. Unwanted heating, however, becomes a possibility with a nonmoving soundhead or excessive wattages. Pulsed ultrasound offers the clinician a technique with less danger of heat buildup (see Pulsed Ultrasound in this chapter). Clinical experience has indicated that a sensation of surface heat with ultrasound should serve as a warning of impending overdosage. A "hot" transducer head not only is uncomfortable for the patient but may be a sign of imminent danger. The head may become overheated from constant and/or repetitive use without adequate time for "cooling off." Despite transmission media, *friction* is still present and must be accounted for. I generally recommend that the clinician apply the soundhead to the palm before placing it on the patient—to ascertain that the head is not too hot.

Interface Heat Buildup

What is more important is that most heat produced by ultrasound is found at an interface where two different tissues abut with a common space intervening. The most prevalent interfaces and the most common sites for heat buildup are the periosteal zones between the hard surface of the bone and the glistening undersurface of the periosteum, where a layer of air, perhaps only one molecule thick, separates bone from periosteum. When the soundwave front enters the tissue and approaches this boundary layer, the change in media from moist tissue to periosteum, to air, and then to bone leads to refraction of the ultrasound wave, multiplied considerably due to the various tissues. When the longitudinal wave-

front is refracted into the periosteal-bone interface layer, the angle of incidence is exaggerated enough to cause reflection from the osseous surface back to the underside of the periosteum and back to the bone again, ad infinitum.

Shearing Effect

The shearing effect transforms the longitudinal wave into a transverse waveform, similar to a Slinky waving up and down rapidly. A transverse wave at 1 MHz produces considerable heat. In that confined space between the bone and periosteum, however, there is little or no opportunity for the heat buildup to dissipate, and a periosteal burn can result. This is an extremely dangerous lesion, since it is not observed by either the patient or the clinician, and the symptoms are not easily noted until systemic manifestations are exhibited (e.g., pain and fever). The transformation of the longitudinal waveform into a transverse waveform at the periosteum is called the *shearing effect* and is to be avoided (Fig. 4-4). In the discussions of techniques, suggestions are made for proper procedures to avoid this danger.

INDICATIONS

Ultrasound is indicated with spasm in neuromuscular and musculoskeletal conditions such as athletic injuries (Figs. 4-5 and 4-6),[3,17,18] scar tissue problems (Fig. 4-7),[19] warts,[20] ganglia (Fig. 4-8), podiatric conditions (Fig. 4-9), and, recently, two conditions involving the female breast: engorgement pain during postpartum nursing[21] and postimplant sclerosis in breast augmentation

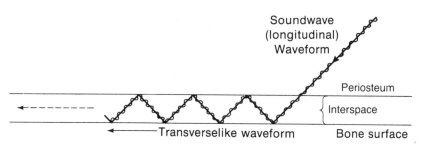

FIG. 4-4. The shearing effect, often leading to a periosteal burn. Note the transverselike waveform of the longitudinal soundwave reverberating between the surface of the bone and the underside of the periosteum.

surgery (F. G. Herhahn, unpublished observations, November 1984). Ultrasound is rarely used for pain management per se. As a phonophoretic procedure, however, it has been found to be of value (see Phonophoresis section).

CONTRAINDICATIONS AND PRECAUTIONS

Contraindications to the use of ultrasound are as follows:

1. Growing epiphyses, the pregnant uterus, bony prominences, ailments of the eye, testicular tissues,[22] and the presence of pacemakers present specific problems and may preclude the use of ultrasound.

2. Special care must be given to patients with sensory losses. When using underwater ultrasound techniques, clinicians should be aware of the possible prolonged exposure of their hands to ultrasound energy.

3. Metallic implants or surgical fixation materials may be a problem when using ultrasound.[10,23] The high-frequency oscillations may disrupt the chemical bonding or cement used with these procedures. Moreover, the interface of metal/tissue may be an ideal site for heat buildup and possible burn.

4. The possible disruption of osteogenic processes in healing fractures suggests extreme caution in use of ultrasound in and around recent fracture sites.[15] Some practitioners, however, claim that ultrasound is a stimulant to osteogenesis.[5] Clinicians should study the

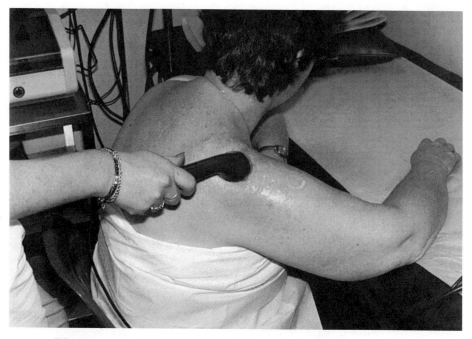

FIG. 4-5. Ultrasound administered for adhesive capsulitis of the shoulder.

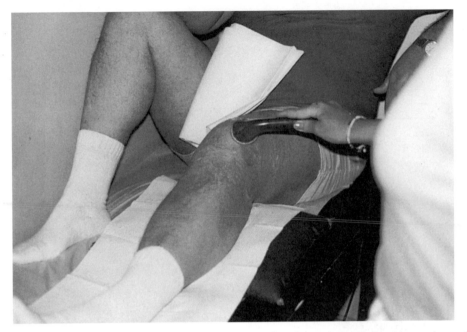

FIG. 4-6. Chronic osteoarthritis in the knee is a suitable target for ultrasound with or without phonophoresis.

professional literature pertaining to this subject carefully before making a decision regarding applications of ultrasound in and around healing fractures.

PARAMETERS

Power Wattage

The recommended dosage for most clinical procedures is 0.5 to 1.0 W/cm^2 of transducer head.[24]

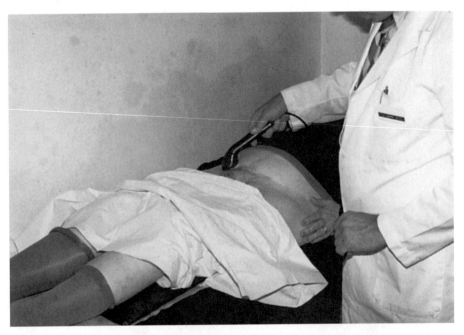

FIG. 4-7. Ultrasound applied to abdominal surgical scars.

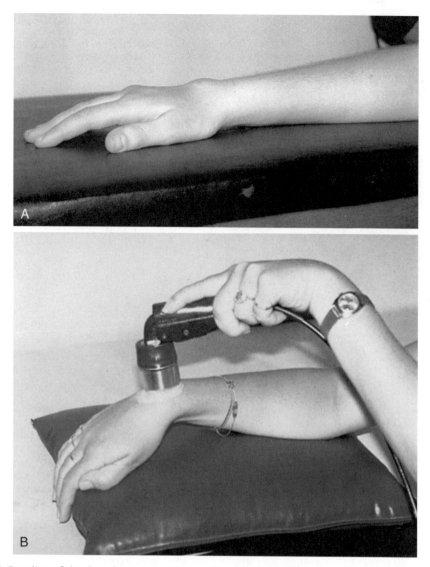

FIG. 4-8. (A) Ganglion of the dorsal wrist. **(B)** Ultrasound applied with hydrocortisone directly to the ganglion, to be followed by shortwave diathermy.

Instruments generally have a crystal area measuring from 5 to 10 cm^2, which allows a total wattage of 2.5 to 10 W—sufficient for most applications, including underwater techniques and phonophoresis. Higher dosages (i.e., 1.5 to 2 W/cm^2) have not proven more efficient; indeed, they have been shown to be less effective than the lower dosages.[24]

and whether surface or subaqueous technique is used. Most manufacturers offer suggested dosage charts with their products. I have found that 2 to 3 minutes of ultrasound treatment is adequate in most cases when administered manually on the surface and that 5 to 8 minutes is adequate when administered under water.

Treatment Duration

Duration of treatment at the recommended dosages ranges from 1 to 8 minutes, depending on the condition treated, the anatomic target site,

PULSED ULTRASOUND

Intermittent or pulsed soundwave propagation is a mechanical device that offers an off/on cycle to

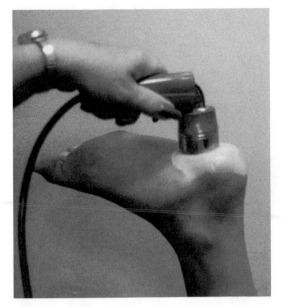

FIG. 4-9. Hydrocortisone phonophoresis administered to the soft tissues surrounding a painful calcaneal spur for its anti-inflammatory effect.

the soundwave flow, usually in the order of 60 pulses per second (Fig. 4-10). This mode, with the transmission of the ultrasound energy off half the time, tends to minimize heat buildup in the target tissues and interfaces.[5] For effective treatment, dosages must be adjusted if the same total wattage is to be administered.

TRANSMISSION MEDIA

Commercial transmission gels vary in consistency, color, and chemistry; all seem to offer effective transmission of the soundwave energy to the patient, however. Mineral oil is used frequently as a successful medium. Water is an excellent conductor of soundwaves.[25]

TREATMENT PROCEDURES

Applying the Soundhead

1. Apply sufficient amount of conduction gel to the target area.
2. Prewarm the transducer head in the palm of your hand or in warm water.
3. After turning on the ultrasound unit to the lowest power setting and allowing warm-up time as suggested by the manufacturer, place the transducer on the target area.
4. Increase power slowly to the desired wattage, measured either in watts per centimeter squared or in total watts.
5. Then move the soundhead slowly over the target zone in a combination of small circular and long stroking motions (Figs. 4-11 and 4-12).
6. Use special care over bony prominences to avoid patient discomfort and periosteal irritation.
7. A timer is usually provided by the manufacturer to ensure proper duration of treatments.
8. Many operational manuals and texts advise sounding appropriate nerve roots following local target zone applications, as determined by dermatome distribution (Table 4-1).

Treatment Regimen

1. Administer ultrasound daily, if indicated.
2. Prolonged treatment without notable progress may indicate that a change is needed in treatment regimen.
3. Administer treatments two or three times weekly.
4. One of the most commonly asked questions regarding ultrasound is whether 20 treatments is the maximum permitted. I know of no such

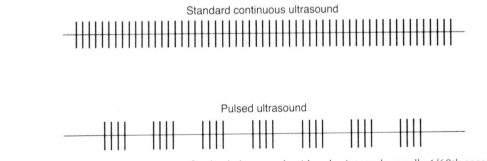

FIG. 4-10. Note the intermittent nature of pulsed ultrasound, with pulse intervals usually 1/60th second. There is rarely any difference in sensation to the patient between the continuous and pulsed waveforms.

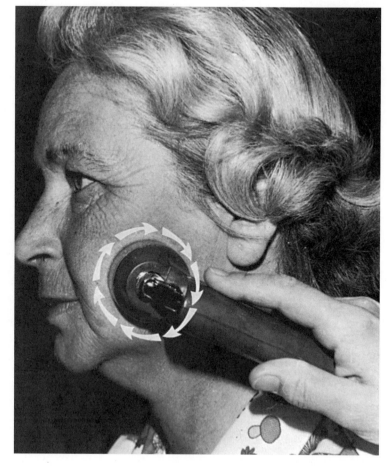

FIG. 4-11. Ultrasound administered to the face and temporomandibular joint area, using a tight circular motion of the soundhead.

limitation and have been unsuccessful in attempting to find such a notation in the literature. I must add, however, that if favorable results have not been achieved in several treatments, another modality or technique should be considered. If the ultrasound is producing continuing favorable effects, a judicious reduction in the number of treatments should be considered to avoid interminable therapy without long-lasting effect. No clinician wants to treat a patient "forever."

Preheating or Pre-Icing the Target

1. Preheating the target area has been reported[26] to retain most of the ultrasound absorption superficially, where circulation is greatest following surface heating.
2. Conversely, pre-icing the target zone tends to drive the circulation deeper and increases the ultrasound absorption at a greater depth.

3. Clinical efficacy of preheating or pre-icing has yet to be determined with a large statistical population.

Subaqueous Technique

1. Fill a horizontal bucket with warm water, deep enough to allow a hand, foot, or elbow to be submerged completely. The receptacle *must* be either plastic or fiberglass; it must not be metal, and a whirlpool must not be used. Manufacturers caution against using the whirlpool tank for several reasons: the inherent danger of ground-fault shocks and the unnecessary dispersion of the soundwave into more than 50 gallons (190 liters) of water when a much smaller quantity will suffice.
2. Place the transducer into the water and turn on the unit for the prescribed dosage, which is identical to the dosage used for manual techniques.
3. Adjust the timing to 5 to 8 minutes.

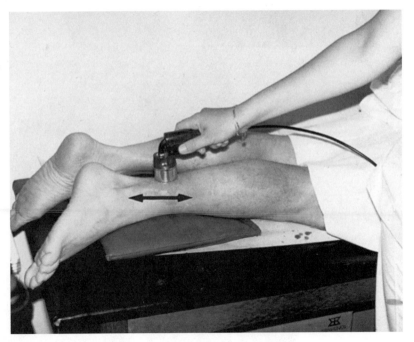

FIG. 4-12. Ultrasound administered as part of a regimen for psoriasis. Following irradiation with ultraviolet, sclerotic patches are sounded with ultrasound for the sclerolytic effect. Recommended dosage: 5 W (total) for 1 to 2 minutes, depending on the extent of the lesion.

4. Place the soundhead facing *away* from the target area, allowing the soundwaves to reverberate off the receptible walls. An increased treatment duration is required because of the obvious attenuation of the intensity due to reflection and dispersion (Fig. 4-13).

5. Some clinicians prefer to direct the soundwaves at the target area, at a distance of 2 to 3 cm from the skin. With this technique, the transducer should be held in the hand, or slowly moved by hand, for the duration of the treatment. A special handle is available for this purpose, minimizing or eliminating the danger of prolonged exposure to the physical therapist.

6. Because success is claimed for both techniques, clinicians are free to select the procedure most practical for their practice.

ULTRASOUND IN CONJUNCTION WITH OTHER MODALITIES

1. If additional modalities are to be administered during the same treatment session, give consideration to the condition of the patient's skin.

2. When using oils or gels, precede the sounding with electrical stimulation or iontophoresis to ensure optimal electrical transmission (see below).

Ultrasound and Electrical Stimulation

1. If electrical stimulation (EMS) is to be combined with ultrasound, the transducer becomes one of the two electrodes in the circuit (Fig. 4-14).

2. Place a secondary electrode on the patient, usually homolateral—neither too distant nor too close and not on an antagonistic muscle. If the upper extremity is to be treated, place the electrode on the brachial area or dorsal spine; if the lower extremity is to be treated, place the electrode on the lumbar spine.

3. The transducer then serves as the stimulating electrode and ultrasound is administered simultaneously.

4. Various forms of stimulating currents are available for this purpose (e.g., alternating, direct, continuous, surged, and pulsed) (Fig. 4-15).

5. Dosages are problematic, since ultrasound may require only a few minutes for sufficient treatment, whereas electrical stimulation may

TABLE 4-1. Ultrasonic Dosage Table*

Condition	Area of Treatment	W/cm²	Minutes	No. of Treatments	Remarks
Abscess	Local or peripheral	0.5–2	2–8	5–10	
Anal fissure	Local and perifocal	1–2.5	4–6	12–15	Under water
Arthrosis	Paravertebral and apophysis	1–3	5–15	10–20	
Large articulations	Local and periarticular	2–3	5–15	10–20	
Small articulations	Local and periarticular	1–2	2–5	8–12	
Bechterew's disease	Paravertebral, apophysis	1–3	5–15	15–30	
Bronchitis bronchiectasis	C3, D1–12†	1–3	5–15	10–25	
Bronchial asthma	Paravertebral C3, D1–12 stellate ganglion from front, paracardiac	0.5–2	5–10	8–15	Frequent sessions, small dosage
Bursitis	Local and peripheral	0.6–1.5	5–10	10–18	In acute conditions minimum dosage
Causalgia	Local, periarterial and perineural	0.5–2	5–10	10–25	
Cellulitis	Local	1–2	5–10	5–15	
Cholecystitis	Local, paravertebral D7–10	1–2.5	5–10	10–20	
Claudication, intermittent	Tibial and along thigh, L4–5, S1–2	0.5–2	5–15	5–20	Avoid excessive dosage
Coccygodynia	L4–S3 paravertebral and local	1–2	5–10	5–12	
Coxitis	Local on every side	1–3	5–12	10–15	
Dupuytren's contracture	Local, along cervical column, paravertebral C5–7	1.5–3	10–15	10–20	
Eczema	Local, periarterial	1–1.5	5–10	5–10	Under water
Endarteritis obliterans	Periarterial	1–2	5–10	5–15	
Epicondylalgia	Local, cervical column, C4	1–3	5–8	5–10	
Erythema nodosum	Local	0.8–2.5	5–10	10–15	
Erythromelalgia	Fem. art.	1–2	5–10	10–25	When treating one side only heavier dosage
Fistula	Local	1–2	5–10	10–20	
Furuncle anthrax	Local, perifocal	0.7–1.5	5–8	8–12	
Hemorrhoid	Local and perifocal	0.8–2	5–6	10–15	Under water
Herpes zoster	Paravertebral and local	2–3	5–10	5–10	
Hypertension, essential	Paravertebral C4–D9	1–2	5–10	5–15	Stop if negative after five treatments
Lumbago	L3–S1 paravertebral and apophysis L2–S4	1–3	10–15	2–10	Daily
Lymphadenitis	Local	0.5–2	5–10	5–15	
Mastitis, puerperal	Local	0.8–2	10–15	10–20	
Myalgia	Local, corresponding paravertebral segments	2–3	5–10	3–20	
Myelitis	At level of medullar focus	0.8–2	5–10	5–10	
Myositis	Local and paravertebral	1–2	5–10	10–20	
Neuralgia, intercostal	Intercostal spaces and paravertebrally	0.6–2	5–10	5–15	Beware of angina pectoris

(Continued)

TABLE 4–1. Ultrasonic Dosage Table* *(Continued)*

Condition	Area of Treatment	W/cm²	Minutes	No. of Treatments	Remarks
Neuralgia neuritis	Local and paravertebral segments	1–3	5–10	5–20	
Osteitis	Local and perifocal	1–2	3–10	10–30	
Osteomyelitis	Local	1–2	5–10	10–30	
Otosclerosis	Mastoid apophysis	0.3–1.5	5–8	5–15	Shave
Parotitis	Local	0.5–2	5–10	10–15	
Periarthritis, humero-scapular	Local and inferior cervical column, stellate ganglion	0.5–3	5–15	10–20	Heavy dosage for local treatment
Periostitis	Local and perifocal	0.8–2.5	3–10	10–30	
Polyarthritis, chronic	Local and paravertebral nerve roots	1–2	5–10	5–15	
Postoperative pain	Local and affected nerve roots	1–2	5–10	4–10	
Prostatitis	Local	1–3	5–10	10–20	
Pruritis	Local	1–2	5–15	15–30	
Pulpitis	Local	0.5–1	3–5	2–4	Through the cheek
Radiculitis	Paravertebral and along nerve	1–2	5–10	5–15	
Raynaud's disease	Local	1–2	3–8	5–12	
Rheumatism	Local and paravertebral segments	0.8–3	5–15	10–30	
Sciatica	L3–S1 paravertebral, then L5–S1	1–2	5–15	5–20	
Scleroderma	Local and related paravertebral segments	1–2	5–15	10–20	
Sclerosis, miliary	Paravertebral	2–3	5–15	10–20	
Sinusitis, frontal	Local	0.3–0.8	4–8	5–10	Through the cheek
Sinusitis, maxillary	Local	1–2	5–10	5–15	
Spondylitis, deformans	Local	2–3	5–10	5–20	
Stumps	Local	2–3	5–15	5–15	
Sudeck's atrophy	Local	1–3	5–10	5–15	Under water
Sudoriparous abscess	Local	1–2	3–5	5–10	Under water
Syndrome scalenus	Local and D4–7	0.6–1.5	5–10	5–10	
Tenovaginitis	Local	1–2	5–10	5–20	
Thromboangiitis obliterans	Local, periarterial and paravertebral along nerve roots	1–2	2–5	20–40	Twenty daily sessions, interval of 30 days, and 20 more treatments
Thrombophlebitis	Local	0.5–2	4–10	5–15	
Thyrotoxicosis	Local	1–2	5–10	5–15	
Tonsilitis, chronic	Maxillar angle	1–2	5–10	10–20	
Tumors	Local	2–3	5–15	10–20	
Ulcus cruris	Local, periarterial, paravertebral	1–3	4–15	5–10	Under water
Ulcus ventriculi	D9–10 front, paravertebral D6–10	1–3	5–10	10–20	

*Composite of European and American reports on use of ultrasound.
†Refer to Fig. 6-3 for dermatome distribution.
Courtesy of Birtcher Corp., El Monte, California.

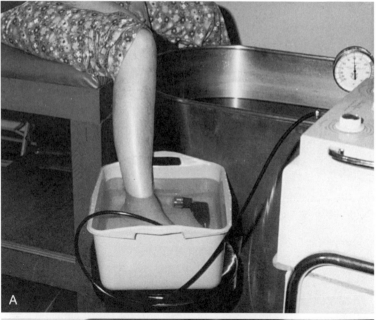

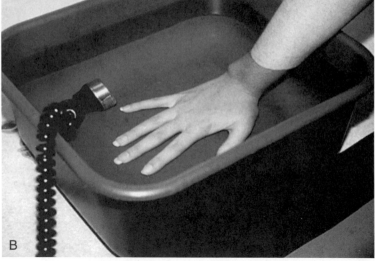

FIG. 4-13. Ultrasound administered under water. The transducer is faced away from the target, necessitating increased exposure to the physical therapist's hands. **A,** Note that the treatment is administered for plantar fascitis in a plastic bucket and *not* in the whirlpool tank (see text). **B,** The targets are post-traumatic scarring and contractures.

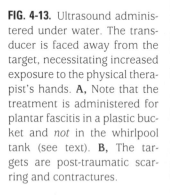

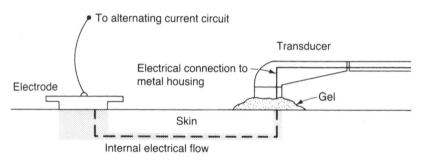

FIG. 4-14. Mechanics of combined ultrasound with electrical stimulation.

WAVE FORM DIAGRAM

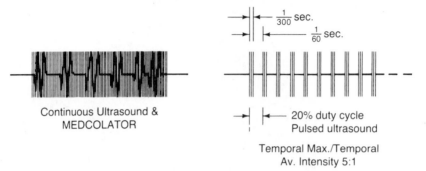

Continuous Ultrasound &
MEDCOLATOR

$\frac{1}{300}$ sec.

$\frac{1}{60}$ sec.

20% duty cycle
Pulsed ultrasound

Temporal Max./Temporal
Av. Intensity 5:1

FIG. 4-15. Waveform diagram of combined electrical stimulation and ultrasound. (Courtesy of Medco Corp., Tinton Falls, New Jersey.)

require 15 to 30 minutes for treatment. A competent clinician will be able to adjust and design a proper dosage combination. Some manufacturers offer units predesigned for combined operations, whereas others make controls (Fig. 4-16).

6. To date, few if any references describe the advantages or disadvantages of the combined techniques. The choice now is clearly the clinician's. Aside from the savings in time and expense, the combined ultrasound and electrical stimulation technique has not been demonstrated to be more effective clinically than when administered separately. Should a breakdown of a combined unit occur, the clinician is left without either. Also, unless two units are available, only one patient may be treated at a time; with separate units, it is possible to use the modalities separately on two patients.

7. I prefer to administer each modality separately, to derive the full benefit of each, with stimulation first followed by relaxation, characteristic of ultrasound.

FIG. 4-16. Ultrasound combined with electrical stimulation from a compact stimulator. One electrode from the electrical stimulation unit is removed so that the lead pin can be connected to the transducer housing, making the soundhead an electrode. (Courtesy of Mentor Corp., Minneapolis, Minnesota, and Birtcher Corp., El Monte, California.)

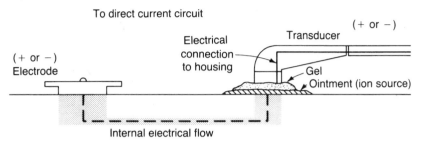

FIG. 4-17. Circuitry for combined ultrasound and iontophoresis. Direction of flow depends on polarity of electrode/transducer and ion used. Electrons flow from negative (−) to positive (+). Traditional terminology lists "current" flow from positive to negative, however.

Ultrasound and Iontophoresis

There have been reports of administration of iontophoresis in combination with use of ultrasound/electrical stimulation equipment.[27] Theoretically, this has not been demonstrated to be possible, since only a continuous direct current will serve for ionic transfer, and time-in-place is a necessary factor with iontophoresis. I know of no available combined units that offer a continuous direct current. Moreover, the movement of the soundhead associated with ultrasound techniques precludes the time-in-place requirement.[28] With this technique, however, there may be phonophoretic rather than iontophoretic effects (Fig. 4-17).

PHONOPHORESIS

The introduction of substances into the body by ultrasonic energy is called *phonophoresis*.[29] This noninvasive procedure has been improperly compared with iontophoresis. Iontophoresis involves the transfer of ions into the tissues, whereas phonophoresis transmits molecules—a different process, although a similar concept.

PHYSICS

Molecules introduced into the target must be broken into component elements and radicals by natural chemical processes and recombined with existing bloodstream radicals. Researchers have claimed penetration of these molecules to depths of 4 to 6 cm[5,30]; however, it is highly unlikely that substances of molecular sizes can be forced through the tissues to those depths. Soundwaves may easily penetrate that deep; however, molecular transfer is another matter.[3] There does not seem to be any clinical evidence of molecular transfer to depths greater than 1 to 2 mm.[29]

TREATMENT PROCEDURES

The technique for phonophoresis is the same as that for standard ultrasound administration.[31-34] The ointments massaged into the target area prior to sounding differ, however (Fig. 4-18). Solutions are not used for phonophoresis, nor is this procedure suited to subaqueous ultrasound. The dissipation of the substances in solutions minimizes molecular transfer, as does the reduction of the soundwave energy when it enters the water.

Molecular Substances

The selection of molecular substances depends on the requirements of the condition—not the disease by name, but the physiologic need of the patient. Chemicals available for phonophoresis at this time are listed on p. 64.

Hydrocortisone

1. Hydrocortisone is available over the counter as a 1 percent ointment.

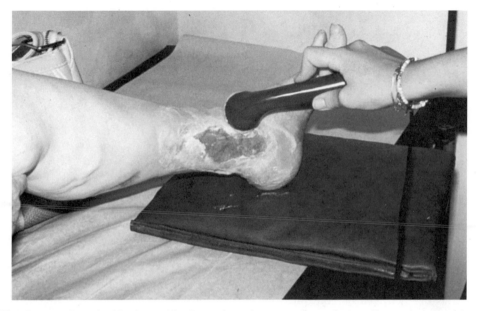

FIG. 4-18. Ultrasound used with zinc oxide phonophoresis over an open lesion. Transmission gel is used in addition to other ointments, since many of the ointments are too viscous for smooth soundhead movements. This procedure follows radiation with the cold laser and/or ultraviolet radiation. (Courtesy of Mettler Corp., Anaheim, California.)

2. Some practitioners prefer the 10 percent ointment.[34] I have not found the higher percentage ointment sufficiently more effective than the lower to be worth the additional expense and difficulty in obtaining the 10 percent ointment.
3. An excellent anti-inflammatory agent, hydrocortisone also provides analgesia in many instances.

Mecholyl

1. Mecholyl can be obtained as an ointment with 0.025 percent methacholine and 10 percent salicylate in a suitable base for either iontophoresis or phonophoresis.
2. Mecholyl is an effective vasodilator and is recommended in vascular conditions, neurovascular deficits, and as a mild analgesic.

Lidocaine

1. Lidocaine is available as a 5 percent ointment (Xylocaine).
2. Lidocaine is used primarily as an analgesic-anesthetic in acute conditions or when decreased sensitivity is desired.

Iodine

1. Available in ointment form, iodine is combined with methyl salicylate (Iodex with methyl salicylate) as an over-the-counter product at local pharmacies.
2. Iodex contains 4.8 percent methyl salicylate, with 4.7 percent iodine in a petrolatum base.
3. Iodine is used as a vasodilating agent, an antiinflammatory agent, and as a sclerolytic agent in cases involving scar tissue, adhesions, calcific deposits, and adhesive joints (e.g., frozen shoulder).

Salicylate

1. Salicylate is available as a 10 percent ointment (Myoflex), over the counter.
2. A basic anti-inflammatory agent, salicylate is also used as a decongestant, as is chemically related aspirin.

Zinc

1. Zinc is available over the counter as a 20 percent zinc oxide ointment.
2. Zinc is a trace element noted for contributing to the healing process (see Fig. 4-18).
3. Zinc is indicated in the treatment of open wounds and lesions (Fig. 4-19).

Mecholyl, xylocaine, Iodex, and Myoflex do not transmit the ultrasoundwave readily. It is highly recommended that these ointments be massaged into the target tissue thoroughly before applying transmission gel and sounding.

Other Products

Other products appliable to phonophoresis are available from manufacturers. Clinicians are advised to check the labeled ingredients for claimed effects with lesser known products.

UNTOWARD REACTIONS AND CONTRAINDICATIONS

Skin and systemic reactions with phonophoresis are rare, but practitioners must be aware of the possibilities and be prepared with antidotes should any of the chemicals listed above present unfavorable reactions in individuals. Keep in mind that allergies and sensitivities to the substance contraindicate its use on the skin as well. For example:

1. Patients who cannot eat seafood should not be treated with iodine. Should skin irritation and itching be reported, the usual antidote is an antihistamine. An alternative chemical should be selected in future treatments.

2. Patients sensitive to metals should not be treated with zinc. These patients usually cannot wear metallic watchbands, jewelry, etc., without having skin reactions and, at times, systemic reactions. Dermatologic consultation should be sought for specific antidotes for the offending metals. Nonmetallic substances should be substituted.

3. If a patient has a reaction to mecholyl with vasomotor shifting, administer a simple stimulant, such as black coffee. Vertigo from orthostatic adjustment is usually momentary.

4. Reactions to hydrocortisone are not as common as patients will have you think. The culprit is usually the chemicals included in the base of the ointment or solution (e.g., novocaine) rather than the steroid itself. Have the patient use an antihistamine skin lotion should any dermal irritation occur.

5. If have not found any untoward reactions to lidocaine.

6. Do not treat a patient with salicylates if he or she is sensitive to aspirin. Seek medical consultation for the specific treatment of symptoms.

It should be noted that, although the above reactions are extremely rare, the efforts taken in the prevention of their occurrence will be well worthwhile.

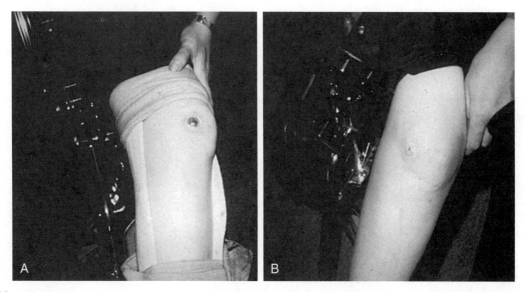

FIG. 4-19. **A,** An open abrasion ulcer on the patella of a wheelchair-bound patient before treatment with zinc oxide phonophoresis. **B,** Enhanced healing and closure following a series of soundings.

RECENT DEVELOPMENTS

Manufacturers have begun to offer interchangeable transducer sizes with each model. The range is usually 1, 5, 7.5, and 10 cm^2. The size differential facilitates sounding anatomic targets of varying dimensions and accessibility. If clinicians limit the variables in their techniques and maintain a *constant total* output, e.g., 5 W, the watts per square centimeter output is automatically adjusted to the soundhead. This only becomes a problem with the smallest soundhead.

Many manufacturers also offer multiple selection for frequencies. The traditional 1 MHz is no longer the only option, 3- and 4-MHz units are now available (*megahertz* refers to the transverse wave alternating current frequency imposed on the crystal, which in turn oscillates at the identical frequency in *pulses per second* of longitudinal sound waves).

When deciding which frequency to use, remember that, in the realm of electrophysics, "the higher the frequency, the deeper the penetration." This is obvious when comparing the long waves and low frequencies of radio transmission with the shorter waves and higher frequencies of x-rays and cosmic and gamma radiation. In the field of sound, however, the reverse holds true. The lower the frequency, the deeper the penetration. Therefore, the 1-MHz transducer will provide deeper effects than the 3- to 4-MHz unit, thus allowing the clinician to select the most effective frequency for the target.

Now physical therapists can choose between transducer head sizes, operating frequencies, continuous vs. pulsed sound, transmission media, manual or subaqueous techniques, digital or analog instrumentation, and combinations of electrical stimulation and among a broad spectrum of chemical substances for phonophoresis . . . making ultrasound one of our most versatile modalities.[35]

Infrasonic Therapy

Unlike traditional ultrasound, infrasound offers oscillations *below* the hearing range, rather than beyond normal (60 to 20,000 Hz) hearing parameters. Current frequencies allotted to ultrasound devices in physical therapy are 3 and 5 MHz. Infrasonic devices operate at 8 to 14 Hz, close to the body's own natural frequency of about 10 Hz. The infrasonic device has been cleared by the FDA as a therapeutic massage modality and is now available to clinicians.

Infrasonic therapy is termed *Qi-Gong* (pronounced "chee-gong"). The advantages of this modality include penetration through clothing, plaster casts, and bulky superficial tissues (Fig. 4-20). No heat is produced, and there is no danger of electrical shocks or burns. The Qi-Gong unit is a deep-massage device, offering the practitioner the benefit of deep, penetrating massage techniques with no discomfort or dangers. It can be safely used in the presence of metallic implants. When the device is placed near water, the ripples of the soundwaves may be seen. The transducer, however, should *not* be placed under water (Fig. 4-21). Contraindications are extremely inflamed skin and the suspicion of blood clots in the calf. The device should not be used in or near water.

FIG. 4-20. An infrasonic massager demonstrating the modality's ability to penetrate clothing for deep massaage. (Courtesy of "Qi-Gong," China Healthways, San Clemente, California.)

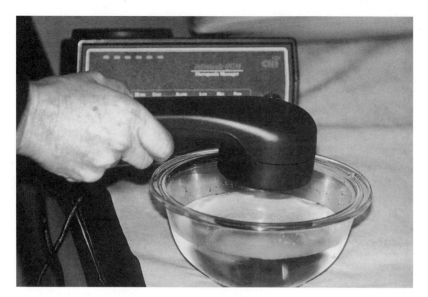

FIG. 4-21. An infrasound device will produce ripples in the water at a distance of 2 to 3 inches (5 to 7.5 cm). (The unit is never placed *in* the water, however.) (Courtesy of GE Miller, Yonkers, New York.)

The physiologic benefits are reported to include increased local blood supply, muscle relaxation, and enhanced healing of wounds and fractures. No gels are required; the treatment soundhead may be placed directly on the skin, or as noted earlier, on clothing, and even 1 inch (2.5 cm) from the skin. Low, medium, and high intensities are programmed for the clinician, and the device has a treatment timer. Lower frequencies are suggested if the patient is weak; high intensities are recommended in the presence of inflammation and edema.

The Qi-Gong concept has been used in China and the Far East for centuries; however, only recently have these sophisticated devices replaced manual techniques used by so-called Qi-Gong masters. Treatment times range from 5 to 20 minutes and may be administered often and in addition to other modalities. The patient experiences a deep, penetrating, and comfortable massage. No undesired side effects have been reported.[36,37]

Additional Information and Pertinent References

Scientific Investigations into Chinese Qi-Gong
China Healthways Institute
117 Avenida Granada
San Clemente, CA 92672

GE Miller, Inc.
45 Saw Mill River Road
Yonkers, NY 10701

REFERENCES

1. Furguson BH: The Ultrasonic Therapy Equipment Standard. US Department of Health and Human Services, Public Health Service, FDA, July 1985
2. Gersten JW: US and extensibility of tendon. Arch Phys Med Rehabil 37:201, 1956
3. Quillen WS: Phonophoresis: a review of the literature and technique. Athletic Training, Summer 1980, p. 109
4. Anonymous: Syllabus. 7th Ed. Burdick Corp., Milton, WI
5. Nyborg WL, Ziskin MC (eds): Biological Effects of Ultrasound. Churchill Livingstone, New York, 1985, p. 9
6. Kahn J: TENS for non-united fractures. JAPTA 62:840, 1982
7. Griffin JE, Touchstone JC: Penetrative and Metabolic Effects of Ultrasonic Energy, p. 122. University of Pennsylvania, Research grant No. RD-2442-M. Department of Health, Education and Welfare, August 1970
8. Madsen PW, Gersten JW: The effect of US on conduction velocity of peripheral nerve. Arch Phys Med Rehabil 42:645, 1961
9. Williams AR: US: Biological Effects and Potential Hazards. Academic Press, New York, 1983, p. 126
10. Reid DC, Cummings GE: Factors in selecting the dosage of US. Physiother Can March 25:1, 1973
11. Anonymous: World Book Encyclopedia. Vol. 8. p. 142. Field Enterprises Educational Corp., Chicago, 1962
12. Dyson M, Suckling J: Stimulation of tissue repair by US. Physiotherapy (London) 64:105, 1978
13. Gersten JW: Non-thermal neuromuscular effects of US. Am J Phys Med 37:235, 1958

14. Behrend HJ, Weiss J: Controversial aspects of US. Phys Med 37:231, 1958

15. Baldes EJ, Herrick JF, Stroebel CF: Biologic effects of US. Am J Phys Med 37:111, 1958

16. Anonymous: Simple Story of Ultrasonics. Birtcher Corp., El Monte, CA, p. 3

17. Quillin WS: Ultrasonic phonophoresis—tips from the training room. Physician Sports Med 10(6): 211, 1982

18. Hiltz D: Calcific Tendinitis. Whirlpool PPS/APTA, 4:1, Spring 1981, p. 10

19. Farber EM, Orenberg EK, Pounds DW (Stanford University): US hyperthermia clears stubborn psoriasis patches for months. Quoted in Medical World News, August 17, 1981, p. 33

20. Contributing authors: Ultrasound and the plantar wart—a review of the subject. Syllabus 42:2, February–March 1966. Burdick Corp., Milton, WI

21. Shellshear M: Therapeutic US in post-partum breast engorgement. Aust J Physiother 27:1, 1981

22. Fahim MS (University of Missouri): Treatment with US blocks spermatogenesis. Quoted in Medical World News, 17:19, May 19, 1976, from a paper presented at the American Fertility Society meeting, Las Vegas

23. Lehmann JF, Brunner GD, Martinis BS, McMillan JA: US effects with surgical metallic implants. Arch Phys Med Rehabil November 1959, p. 483

24. Reid DC, Cummings GE: Factors in selecting the dosage of US. Physiother Can 25:1, 1973

25. Griffin JE: Transmissiveness of US through tap water, glycerin and mineral oil. Phys Ther 60: 1010, 1980

26. Roubal PJ: US—clinical observations. Council Licensed PT NY Tech J 5:28, 1977

27. Smith K: Physical Therapy Department, Memorial Hospital, Amsterdam, NY, 1975

28. Kahn J: Clinical Electrotherapy. 4th Ed. J Kahn, Syosset, NY, 1985, p. 6

29. Antich TJ: Phonophoresis—the principles of the ultrasonic driving force and efficacy in treatment of common orthopedic diagnoses. JOSPT 4:99, 1982

30. William AR: US—Biological Effects and Potential Hazards. Academic Press, New York, 1983

31. Kahn J: TMJ pain control. Whirlpool, PPS/APTA, Fall 1982, p. 14

32. Borden RA: Phonophoresis using cortisol (hydrocortisone). Bull SC PT Assoc 11:1, 1978

33. Dyson M: Therapeutic applications of ultrasound. p. 121. In Nybord WL, Ziskin MC (eds): Biological Effects of Ultrasound. Chuchill Livingstone, New York, 1985

34. Kleinkort JA, Wood F: Phonophoresis with 1% vs 10% hydrocortisone. Phys Ther 55:1320, 1975

35. Hayes KW: The use of ultrasound to decrease pain and improve mobility. Clin Rev Phys Rehabil Med 3:271, 1992

36. Feeney T: Ultrasound: to heat or heal. PT Advance March 13, 1995, 28-29

37. Byl NN: The use of ultrasound as an enhancer for transcutaneous drug delivery: phonophoresis. Phys Ther 75(6):539-553, 1955

5

Electrical Stimulation

One of the oldest and most effective modalities used in physical therapy is electrical stimulation. From the ancient Greek technique of electric eels placed in a footbath, through Benjamin Franklin's approach to his neighbor's frozen shoulder utilizing rudimentary batteries for a source of power, to today's broad spectrum of equipment, electrical stimulation has had and still has a respected place among the standard modalities of the physical therapist. The wide variety of stimulation apparatus all have in common a single purpose: the stimulation of tissues for therapeutic responses. Muscles are stimulated to relax or contract; nerves are stimulated to produce analgesia or to increase dormant motion; bone is stimulated to enhance growth; and general circulation benefits from the stimulation of all these tissues. No matter what the brand name, model, or design, the end result is the same.

The terminology of electrical stimulation is at best confusing.[1] *Nerve* stimulation has been differentiated from *muscle* stimulation for years in every text written on the topic. This differentiation, however, is very difficult to observe clinically, and much of the confusion is justified. I believe that all stimulation is neurologic, whether motor, sensory, or proprioceptive, since the neurofibrils within muscle tissue are true communicators of the stimulation current to the muscle cells, involving the motor point, if intact, or not, as with denervation. References can be found relating to the direct stimulation of muscle, although at such high intensities as to make this procedure clinically unacceptable.[2] Recently, a new term was introduced to practitioners: *functional electrical stimulation* (FES), apparently differentiating this phenomenon from other types of stimulation. No one has yet explained this difference adequately, nor have clinicians, to my knowledge, found any reason for the new term. Therefore, we now have *neuromuscular stimula-*

tion (NMS), *electrical muscle stimulation* (EMS), *functional electrical stimulation* (FES), and *transcutaneous electrical nerve stimulation* for pain control (TENS).

The significance of the above is clearly illustrated by the rulings of the Medicare administration: coverage for electrical muscle stimulators is dependent on the presence or absence of the associated nerve ("intact nerve supply"). This adds to the general confusion, since denervated muscles may demand stimulation to a greater extent than will normally innervated musculature. The section on electrophysiology of the American Physical Therapy Association (APTA) developed a comprehensive reference for terminology in electrotherapy. It is highly recommended for clinicians, academicians, students, and manufacturers.

PHYSIOLOGIC RESPONSES

Physiologic responses to EMS include

1. Relaxation of spasm
2. Monitored contractions of muscles, simulating active exercise
3. Increased production of endorphins (believed to be a consequence of electrical stimulation. This natural, body-generated analgesic is produced normally when the body detects a painful stimulus. Researchers have found that the body may be fooled into increased production of endorphins by nonpainful electrical stimulation.[3–5]
4. Relatively weight-free exercises, depending on patient position and electrode placements
5. Increased fiber recruitment, since most, if not all, fibers will respond to stimulation, differing from normal, active motion, which may recruit only a percentage of fibers

6. Circulatory stimulation by the "pumping action" of the contracting musculature
7. Enhancement of reticuloendothelial response to clear away waste products

INDICATIONS

Electrical stimulation is indicated wherever the above physiologic responses are desired. Most often, electrical stimulation is employed to provide exercise patterns when patients are unable to perform them due to pain, restriction in range of motion, or other dysfunctions of the neuromuscular system. Electrical stimulation is not limited, therefore, to the musculoskeletal system, but may be utilized in gynecologic, urologic, and ocular musculature problems and, most recently, in temporomandibular joint (TMJ) and other dental problems.

CONTRAINDICATIONS

The patient's general health, as well as the specific diagnosis, will determine the advisability of electrical stimulation. The presence of the following conditions would preclude using electrical stimulation as a treatment modality:

1. Fresh fractures, to avoid unwanted motion
2. Active hemorrhage
3. Phlebitis
4. Demand-type pacemaker—newer types may suggest extreme caution, rather than prohibition

EQUIPMENT
Generators
General Considerations

Electrical generators of many types are utilized for clinical electrical stimulation. No evidence has been offered to indicate that any one model, brand, or system is suited to all cases and personal preferences of practitioners. The clinician's choice will depend to a great measure upon his or her clinical experience with each, rather than on the manufacturers' claims. Clinical techniques for various units are listed below.

Types of Generators

1. Traditional low-voltage current, less than 100 V, under 1 kHz (Figs. 5-1 and 5-2).
2. High-voltage direct current, utilizing the extremely short duration pulse (microseconds) theoretically to increase penetration, generally in the range of 300 to 500 V (Fig. 5-3).

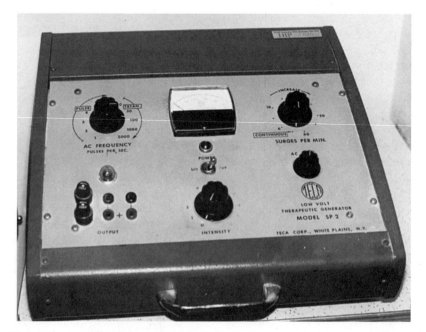

FIG. 5-1. Traditional low-voltage generator showing direct current and alternating current circuits. (SP-2 Model, courtesy of TECA Corp., Pleasantville, New York.)

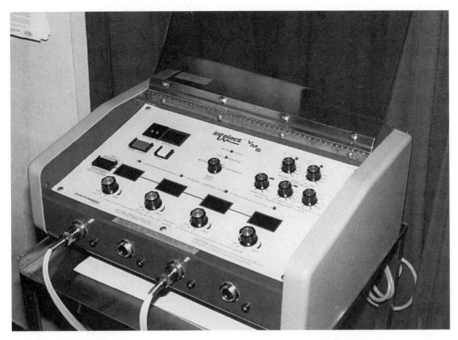

FIG. 5-2. Updated low-voltage generator with multiple-parameter controls. (Intelect VMS, courtesy of Chattanooga Corp., Chattanooga, Tennessee.)

3. Interferential currents in the range of 4,000 to 4,100 Hz, with a net frequency in the interference zone of 80 to 100 Hz. Power is in the low-voltage range (Fig. 5-4).

4. Compact units, much like TENS apparatus, providing parameter selection in the frequency range of 1 to 120 Hz, pulse widths from 50 to 400 μs, ramping (rate of rise)

FIG. 5-3. Typical high-voltage direct current generator. (Courtesy of Dynamax Corp., Preston Corp., Clifton, New Jersey.)

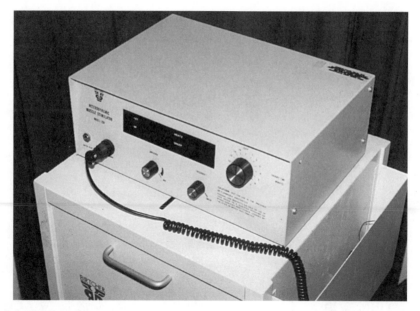

FIG. 5-4. Interferential current generator. (Courtesy of Birtcher Corp., El Monte, California.)

times, and amplitudes in the milliampere range (Fig. 5-5).

5. TENS. Although TENS technically are electrical stimulation units, they are designed specifically for pain control rather than muscle stimulation. Parameters are variable, targeted to stimulate A fibers primarily, utilizing frequencies in the range of 1 to 120 Hz, pulse widths from 50 to 300 μs, with medium range amplitudes in the range of 10 to 50 mA. Waveforms, however, differ somewhat from electrical stimulation units, and electrode placements, treatment protocols, and techniques are considerably different (see Ch. 6) (Fig. 5-6).

6. Medium-frequency generators, designed to utilize specific frequencies in the range of 2,400 to 2,500 Hz ("Russian" type). These are effective in athletic injury management.[6]

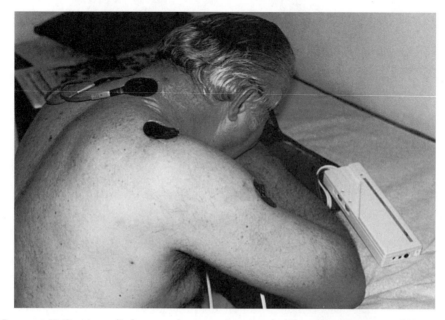

FIG. 5-5. Compact EMS unit applied across the trapezius muscle bilaterally. (Courtesy of Neurotech Corp., Boston, Massachusetts.)

FIG. 5-6. Partial array of TENS and EMS units available to clinicians.

7. Subliminal generators, utilizing stimulation at a nonsensory level to pinpoint targets (i.e., trigger or acupuncture points) with micro-ampere intensities. Interest in microampere electrical stimulation is growing rapidly. Still needed are more clinical evaluations and controlled studies to validate current claims.
8. Programmable units, with variable parameter selections and monitoring facilities. These permit home units to be "programmed" by the clinician and used at home by the patient, with parameters identical to those used in the clinic (Fig. 5-7).

Work done in connection with the Neurotech unit in England has indicated the ability to transform "slow twitch" fibers into "fast twitch" fibers by stimulation with high-frequency currents, and to transform "fast twitch" fibers into "slow twitch" fibers with lower frequency stimulation (R. Jones, personal communication). Additional research and development of specialized parameters and techniques with this equipment in the treatment of old and resistant Bell's palsy has also been reported from England (D. Farragher, personal communication). A third application of programmed stimulation, with

FIG. 5-7. Programmable EMS unit, showing portable module in data-transfer receptacle. (Courtesy of Neurotech Corp., North Andover, Massachusetts.)

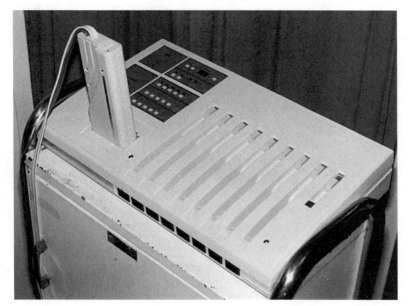

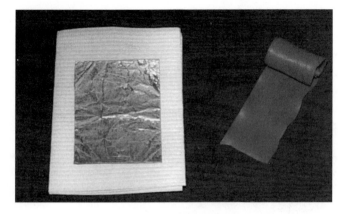

FIG. 5-8. Fabricated electrode, using two soft paper towels, superimposed with folding layers of household heavy-duty aluminum foil cut to size; a soft rubber bandage is used to secure electrodes in position.

specific parameters and protocols with this unit, has been the stimulation of spastic paraplegics to obtain active motion (N. Shindo, personal communication).

Electrodes

The clinician may choose from a large variety of electrodes, some of which are listed below.

1. Commercial pads of rubber-backed canvas or felt have a variety of connecting mechanisms.
2. Moistened paper towels, with aluminum foil plates, necessitate "alligator" clips for connection to leads (Fig. 5-8).
3. "Sponge" types have insert electrodes with rubber carriers.
4. Carbonized rubber TENS-type electrodes use transmission gel.
5. Copper-tipped electrodes are used for internal administration (e.g., intranasal, intravaginal).
6. The physical therapist uses a fingertip, when necessary, for children and animals (Fig. 5-9). This technique involves placing one electrode on the patient, at a distance from the target area, and another electrode on the physical therapist's forearm. A small amount of transmission gel on the tip of the physical therapist's finger offers a small area of stimulation with a familiar instrument to an apprehensive child or, in veterinary practice, a nervous animal.

The fabrication of electrodes is illustrated in Figs. 7-14 to 7-27.

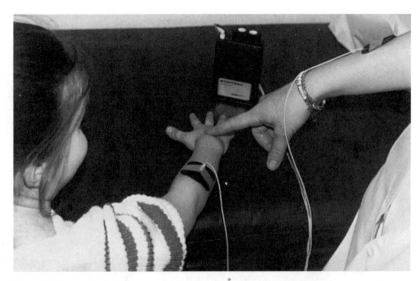

FIG. 5-9. The physical therapist's finger serves as an electrode. Standard electrodes are placed on the forearm of both the patient and the physical therapist. A drop of transmission gel applied to the therapist's fingertip completes the circuit for stimulation. (Stimtrac, courtesy of Stimtech Corp., Randolph, Massachusetts.)

Electrode Dimensions

Electrode sizes will vary, depending on the treatment technique selected and the current configuration desired.

1. Equal sizes for equal distribution of current[7]
2. Differential sizes for current-shaping
3. Special instruments for internal application, (e.g., nasal, vaginal)

Electrode Tips

All electrode tips (Fig. 5-10) are easily attached to the ends of standard lead wires and may be obtained from electronic suppliers.

1. Alligator, for aluminum plate connections
2. Banana, for standard receptacles
3. Telephone, for pin receptors
4. Snap, for button-type connections

Security Devices

Various devices are required to prevent movement of electrodes during treatment. Care should be taken to avoid using materials that, when wet, will conduct current and lead to increased voltages and possible burns. Securing devices include

1. Soft rubber bandages, 1.5 by 60 inches (4 by 150 cm) cut from larger rolls
2. Lightweight sandbags, 1 to 2 pounds (0.5 to 1 kg) only
3. Adhesive tapes

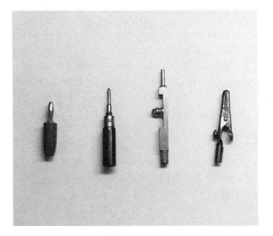

FIG. 5-10. Electrode tips: banana, telephone, Macintosh, and alligator types.

4. Velcro bands
5. Adhesive gels

CURRENTS

There are several general current types and modes used in electrotherapy (Fig. 5-11).

Alternating Current

Alternating current provides alternating polarity, i.e., changes from positive to negative and vice versa, many times per second, too rapidly to offer any polar effects.

When the alternating phases are smooth and equal in energy levels, they are often referred to as *sinewaves,* since the shape of the current resembles the traditional form of the mathematical sinusoidal curve. This form of current is generally applied to neuromuscular components with no reaction of degeneration (RD) to provide relaxation to muscles in spasm or to exercise weak, atrophic, or debilitated musculature. It is a comfortable waveform, easily controlled and modified as needed, with no chemical characteristics at the electrode surfaces.

Direct or Galvanic Current

Direct current represents a constant electron flow from the negative electrode to the positive with no oscillations or alternations. Polarity remains constant, as predetermined by the clinician.

Continuous Direct Current

Continuous direct current is used primarily for iontophoresis.

Interrupted Direct, or Pulsed Galvanic, Current

Interrupted direct current or "pulsed galvanic" current is the preferred modality for stimulation of neuromuscular components with RD, since the body loses the ability to respond to alternating current in the presence of RD. It is often used to stimulate normally innervated musculature; however, this form of current has chemical components that tend to cause skin irritation if left in place too long or if the intensity is too high. These chemical manifestations of direct current are a great detriment, since they account for many of the past problems with direct current. Still, the chemical effects of direct current are sought in

 Continuous direct current

 Surged direct current

 Interrupted direct current (pulsed)

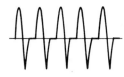 Continuous alternating current

 Surged alternating current

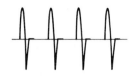 Interrupted alternating current (pulsed)

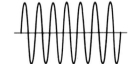 Continuous faradic

Surged faradic

Interrupted faradic (pulsed)

FIG. 5-11. General current types and modes.

the treatment of several conditions, particularly hyperhidrosis (excessive sweating), because of their therapeutic effects.[8]

Surged Direct Current

Surged direct current is not used today because the slow rise of the wave front lends itself to accommodation by the tissues, resulting in minimal or no contractions. (Accommodation occurs when the stimulating current waveform peaks too gradually to cause muscular contraction or flows continuously without modulation or variation for extended periods.) One manufacturer offers a surged direct current mode consisting of a sawtooth waveform of monophasic current, minimizing accommodation (Amrex Corp., Hawthorne, CA).

Chemical Effects

Positive Pole

The anode, or the electrode representing the lesser concentration of electrons (−),

1. Is acidic—hydrochloric, oxygen-rich
2. Is sclerotic (toughening agent)
3. Repels positive (+) ions
4. Decreases irritability level (analgesic)

Negative Pole

The cathode, or the electrode representing the greater concentration of electrons (−),

1. Is alkali—sodium hydroxide ("lye"), hydrogen-rich
2. Is sclerolytic (softening agent)
3. Repels negative (−) ions
4. Increases irritability level (stimulating)

Faradic Current

Faradic current is found less frequently in American-made equipment than in European apparatus. Foreign units tend to favor this composite current form, which is derived from secondary coils and presents an effective although usually irritating stimulant to the patient. Response to faradic current is also lost in the presence of RD. It resembles alternating current in most of its characteristics, with the irritation level being somewhat greater, almost as high as with direct current; thus, students commonly use the phrase "acts like alternating current, feels like direct current."

MODES

The three forms of current now in use can be administered in three major modes.

Continuous Mode

Alternating current at any frequency, as long as there is no break or variation in the current flow, may be used. If the rate (frequency) is greater than 50 Hz, the current is deemed to be *tetanizing*, i.e., producing a constant contraction of normally innervated musculature. This mode is used to obtain relaxation with muscles in spasm.

Direct current (straight-line galvanic current) has only one use—iontophoresis. Faradic current is the same as alternating current.

Surged Mode

Alternating current may be surged to stimulate normal muscle contraction for exercise purposes. In the surged mode, peak intensity is reached in microseconds or milliseconds and builds to maximum with relative slowness. Slow surges (i.e., 5 to 10 per minute) are thought to stimulate "slow fibers" predominantly.[9]

Direct current is no longer used in the surged mode because of accommodation.

Faradic current is the same as alternating current.

Interrupted or Pulsed Mode

Alternating current, when interrupted sharply, reaches peak intensity immediately, causing a brisk response in the muscles. This mode is thought to stimulate the "fast fibers."[9] If interruptions are higher than the rate of 50 per second, tetanic contraction may occur. Interrupted alternating current is often used in management of athletic injury.

Direct current, or interrupted galvanic current, is the mode preferred for management of damaged neuromuscular components, i.e., in the presence of RD. The clinician must be careful not to use currents with rates that rise too rapidly, since denervated muscles lose the ability to accommodate and may actually respond better to a slowly rising waveform. With interrupted direct current as the stimulator, it is recommended that the stimulating electrode be maintained as the *negative*. It is more irritating than the positive (see p. 122) and would therefore require less voltage to produce effective results.

WAVEFORMS

Whether the modes are continuous, surged, or interrupted, alternating current must have form, or waveshape (Fig. 5-12). Direct current also must conform to one of several shapes, for classification purposes as well as differential physiologic effects.

Sinewave

The term *sinewave* describes the alternating current found in many modern units, in both contracted and expanded forms. Sinewave usually exhibits equal energy levels under the positive and negative phases.

Rectangular

Sometimes referred to as a *square wave,* the rectangular form of current is usually descriptive of a direct current with a rapid instantaneous rise, prolonged duration, and sharp drop-off. When the duration equals the intensity (graphically), the term *square wave* is used.

Spike Wave

The *spike-wave current* features a tentlike appearance in which the rate of rise is rapid, but not instantaneous, falling back to zero rapidly immediately upon reaching maximum.

Combined Waves

Combined forms are found in many models of generators, offering clinicians combined effects.

Twin-Spiked Forms (High-Voltage Pulsed Direct Current)

In the attempts to penetrate deeper into the tissues with the stimulation, the adage of "the higher the frequency, the deeper the penetration" took on a new form: "the shorter the pulse width, the deeper the penetration," since Pulse

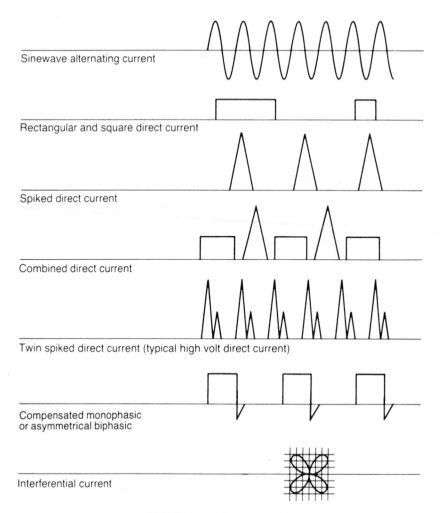

FIG. 5-12. Waveforms in EMS.

width × Frequency = C (speed of light, 300,000 km/s). However, a second phenomenon became a factor: the strength/duration curve (see Fig. 8-5). The shorter duration pulses necessitated considerably higher voltage than with traditional low-voltage procedures. Therefore, *high voltage* apparatus requires voltages in the 300 to 500 V range to drive the 7 to 20 μs pulses. These spike waves were then twinned to reduce skin irritation; the second or "echo" wave helped to "round out" the sharper first spike when modified by skin resistance.

Techniques with high-voltage pulsed direct current (HVPDC) were much the same as with standard procedures; muscle contractions were obtained at somewhat deeper levels than with low-voltage equipment. The question of relative comfort between low- and high-voltage stimula-

tion depends mostly on the clinicians' techniques and patients' individual sensitivities. Those patients sensitive to *voltages* will favor low-voltage techniques; those with sensitivities to *frequencies* would probably favor the high-voltage procedures. The higher frequency (shorter duration) pulses penetrate deeper, bypassing the superficial nerve endings and offering somewhat more comfortable treatments, provided that the patient is not unusually sensitive to the higher voltages.

HVPDC equipment also offers clinicians alternating modes as standard fare, facilitating reciprocal techniques. The direct current (galvanic) mode also provides choice of polarities at the electrodes. Recommended selectivity for polarity ranges widely between manufacturers and practitioners. *Positive* for pain is a traditional rule, with *negative* for stimulation. *Alternating* polarity is suggested when the target is open-lesion healing. The *anodal* acid (HCl) bias for sclerotic effects, coupled with the *cathodal* alkaline bias (NaOH) sclerolytic effects, work well together, but not simultaneously. Treatment protocols often call for changing polarity at the active electrodes periodically during treatment sessions and/or with alternate sessions.[10]

Recent additions to the HVPDC equipment include ramped waveforms, multiple electrodes, and individual controls for each circuit. In most instances, *three* electrodes are utilized for treatment. The two "active" pads with identical polarity are placed appropriately over the target area(s), while the larger dispersal pad, with opposite polarity, is placed at a distant point (Fig. 5-13). A problem may develop when the dispersal pad is the *anode* and the two smaller pads are the *cathodes*. This is contrary to the rule that the cathode should *always* be larger than the anode in order to decrease current *density* at the alkaline (NaOH) interface. However, the extremely low microamperage generated with the short-duration, twin-spiked waveforms minimizes the electrochemical effects, providing enough polarity for selective wound healing, but not for skin

irritation. (It should be noted and emphasized that HVPDC is *not* suited for iontophoresis due to the pulsed nature of the waveform and low microamperage; see Ch. 7.)

More than 60 years ago, a form of high-voltage current was utilized by the early physical therapists. This modality provided actual sparking between the generator and the skin, but nevertheless served as a forerunner of modern high-voltage equipment. It is interesting to see how many of our "new" concepts are really quite old but dressed up in new clothes, titles, and terminology (Fig. 5-14).

Compensated Waveforms

Many waveforms are termed *compensated* because of the attempt to reduce the chemical effects of direct current polarity, by providing opposing polarity phases in the waveform to cancel the bias of the predominant phase.

Interferential Current

Interferential currents[11,12] are unusual in that they combine two high-frequency waveforms

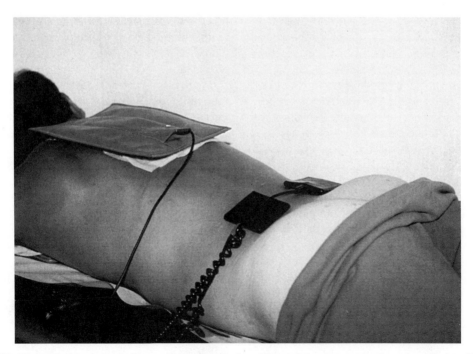

FIG. 5-13. High-voltage, pulsed direct current applied to bilateral lumbar musculature, with the dispersal electrode placed over the dorsal spine. Note the use of a moist towel under the dispersal pad to ensure good skin contact. Securing sandbags (removed for the photo) would eliminate the air-spacing seen here.

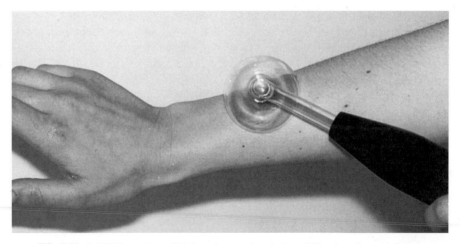

FIG. 5-14. A 1926 version of high-voltage stimulation. Effective, but outdated.

(e.g., 4,000 and 4,100 Hz) in a crossed pattern, so that the net frequency resulting from cancellation/reinforcement phenomena at or near the crossing point amounts to 100 Hz. The traditional popularity of 100 Hz as a stimulating frequency is based on the proximity of this rate to the normal tetanizing frequency of 50 to 100 Hz. Tetanizing frequency refers to the rate of stimulation required to produce a smooth, sustained contraction. With RD, however, this figure drops to lower rates (e.g., 10 to 20 Hz) and can be diagnostic. The concept of interferential currents, therefore, relates to the penetrating quality of the higher frequencies/shorter pulse widths in reaching deeper tissues. Because the higher frequencies do not have the ability to produce contractions due to the shorter pulse durations, and the 100-Hz frequency does have this ability, the units are designed to produce this optimal frequency (100 Hz) at depths greater than is possible with direct surface electrodes utilizing 100 Hz as the operating current, as does traditional low-voltage apparatus. Interferential currents do not usually produce visible contractions unless amplitudes are set high; however, the patient will report contraction sensations at lower intensities.

It may be noted that the presence of nerve damage (RD) will be indicated by greater reaction/response and/or sensation with the *lower* frequencies (i.e., 1 to 10 Hz when in the "sweep" mode). A suggested reason for this phenomenon is the wider pulse widths in the low-frequency bands, more favorable to damaged neural tissues. With more extensive neuropathy, however, the biphasic (alternating current) medium frequencies associated with interferential currents may *not* elicit contractions at all.

Russian Stimulation

"Russian" stimulation consists of frequencies in the 2,400 to 2,500 Hz band, pulsed at various rates. Operating in the higher voltage ranges, 100 to 500 V, this modality has been reported to be effective in the management of athletic-type injuries. Whether this is due to the traditional high discomfort thresholds of athletes or to the physiologic benefits of high-voltage/medium-frequency stimulation remains to be determined.

Diadynamic currents have been utilized in Europe for many years. Only a few models were introduced in the United States in any significant numbers during this time. To the best of my knowledge, there are no American products in this field. Most of my own exposure to this equipment has been at the Stony Brook University laboratory, where equipment from Enrauf-Nonius in Delft, Holland, was available. I was fortunate, however, to attend several conferences in the United States and Canada, where this modality was discussed in depth by "Sandy" Rennie, Professor at the University of Alberta, Edmonton, Canada.

The basic electrophysics of this waveform involves single- or double-phase rectification of an alternating current. By administering these waveforms in a continuous mode, or in combined forms, modulations are obtained that produce varied sensory and motor effects. Analgesia and muscle stimulation are primary goals, utilizing several forms of diadynamic waveforms, generally in the milliampere range. Polarity is a factor in determining treatment parameters and techniques, with differentiation of effects between anodal and cathodal charges. For the actual treatment procedures and parametric controls the reader is referred to the manufacturers' operating manuals, since I am not able to report, with accuracy, any clinical details at this time (Fig. 5-15).[13]

MICROAMPERE ELECTRICAL STIMULATION

No longer the "new kid on the block," microampere electrical stimulation (MES) is still undergoing the rigors of general acceptance. Claims by manufacturers, dealers, and users far exceed documented research. Many endorse-

ments come from athletic trainers and similar sources dealing with sports-type injuries and management. (Our patients are far from athletic types!) Some of the references available appear to be endorsements for specific brands of MES apparatus.

Background

Based on the work of early investigators in this field, the electronic nature of human physiology is explored in terms of intensity levels, polarity, and waveforms. Research results involving the effects of electrical stimulation in a variety of pathologies have been extrapolated for application in a *general* sense. The specific work of Nordenstrom[14] is singular in its relevance; however, it does not apply directly to the concepts of MES as a therapeutic device. Some references of Picker[15] and Gersh[16] indicate electrophysical possibilities in the therapeutic area.

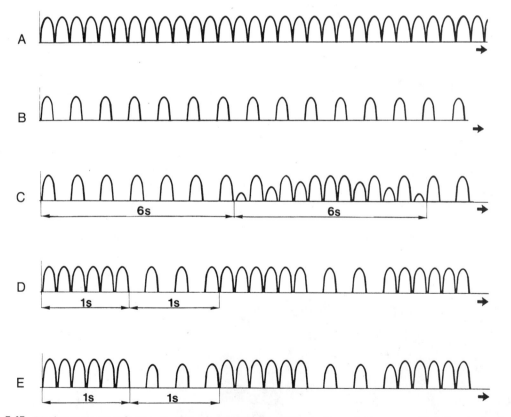

FIG. 5-15. Diadynamic waveforms. **A,** DF—double-phase rectified alternating current. **B,** MF—single-phase rectified alternating current. **C,** LP—slow alteration between MF and DF. **D,** CP—rapid alternation between DF and MF. **E,** CPid—similar to CP, but the amplitude is increased by approximately 10 percent during the DF phase. (Courtesy of Enrauf-Nonius, Delft, the Netherlands.)

Rationale

The Arndt-Schultz law mentioned elsewhere in this book offers the theory that "weak stimuli increase physiological actions, while very strong stimuli may inhibit or prevent activity." MES certainly follows the precepts of this law. Picker[15] stated it very well when he commented on the possible advantage of "whispering to the body with microcurrents instead of shouting with milli-amps."

Comparisons with TENS, low-voltage, and high-voltage stimulation procedures have been unsuccessful in establishing the clear-cut validity of the superiority of MES. Low-voltage direct current at microampere levels and high-voltage pulsed direct current have both been reported successful in enhancement of tissue repair.[10,17] The HeNe laser[18] offers subliminal, MES-type stimulation, along with the laser radiation.

Technical Data

Microampere stimulation, therefore, in the relatively lower voltage range (60 V), apparently is successful in reducing pain and enhancing tissue healing. The availability of polarity selection to the clinician is an important factor. With the units discussed here (the "My-o-matic," Monad Corp.; and the Dynatronics 200) (Fig. 5-16), the clinician is offered the choice of positive, negative, or bipolar stimulation. Positive was recommended for pain control, the negative for sclerolysis or stimulation, and the bipolar or alternating as a catch-all or terminating mode for all treatments. My experience has been favorable regarding the concept of "positive for pain" based on the slightly acidic reaction and oxygen-rich environ-ment of the anode. The highly irritating alkaline NaOH produced at the cathode is appropriate for sclerolysis and stimulation. The manufacturer's suggested bipolar phase at the termination of each treatment may be advised in an attempt to leave the tissues in a state conducive to neutrality.

The various phases available to the clinician, such as the search, treatment, and interferential modes, are best left to the operation manuals with each unit, since manufacturers provide unique and specialized parameter controls for obvious reasons.

It is the *concept* of MES therapy that needs further study and evaluation. Based on the work of investigators and clinicians, this innovative form of subliminal electrical stimulation warrants the attention of clinicians, academicians, and researchers.

According to the work of Nordenstrom[14] and Becker et al.,[19] there is a "second circuit" of electrical transmission within the human body. Charged ions and radicals travel via the circulatory capillary system, mimicking the neural system with its axon-dendrite transmission. This "second circuit" provides an adequate basis for electrical transmission within the body. Unlike the neural system, however, this second circuit is modulated by several factors specific to itself (i.e., polarity and intrinsic resistance patterns). Pathologic tissues offer *increased electrical resistance* to incoming ionic transmission via capillary transport. Since the ions and radicals in transmission are "charged," the unusually high positively charged resistance prevents nutrients and other vital substances from entering the pathologic area to effect repair. Thus the body must find alternate

FIG. 5-16. The control panel of the Myomatic MES apparatus. A dual-channel microampere generator with frequency and pulse duration selection availabilities. (Courtesy of Monad Corp., Pomona, California.)

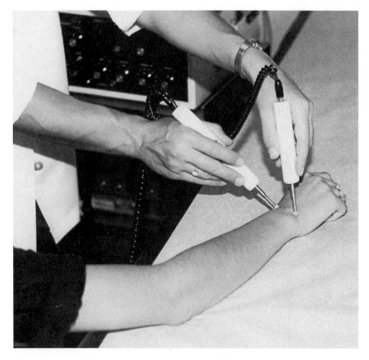

FIG. 5-17. MES probes applied for wrist pain using moistened cotton swabs. Controls are in a thumb button on the main *(left)* probe. (Courtesy of Monad Corp., Pomona, California.)

routes to the damaged areas to enhance favorable metabolism and repair. This is a time-consuming procedure, costly to rehabilitation.

The administration of low-intensity (i.e., microamperage) currents serves to "boost" the electromotive force behind the moving ions and radicals sufficient to allow entry into the high-resistance zones of pathology so that favorable metabolism and repair can take place. Therefore, in essence, MES serves to enhance healing by stimulation to the *circulatory* system, rather than the *neural* system, negating the "N" in "MENS."

The polar effects favor release of oxygen at the anode, which apparently favors analgesia, as well as healing. Thus, MES is also useful in providing pain relief in many instances.

Techniques

Administration techniques, as with most modalities, vary. Commentaries on the specific procedures recommended by manufacturers are best found in their operating manuals. These techniques include use of probes and standard electrodes.

1. Probes with moistened cotton swabs are held by the clinician on both sides of the target area, so that current flows through or along the painful region (Fig. 5-17).

2. Standard (TENS-type) electrodes, with appropriate conductive gel, are placed astride the target tissue (Figs. 5-18 and 5-19).

3. Techniques involving the use of "roller" cylinders are recommended when larger muscle groups are the targets.

4. Stretching shortened muscle groups or compression of extended groups may be accomplished by recommended manual techniques with the electrodes.

5. Targeting of motor points, acupuncture points, and intratendinous structures (i.e., Golgi tendon organs) determines the planned effects of stimulation, analgesia, or relaxation.

6. Electromassage techniques require the clinicians to place electrodes on their body, as well as on the patient's body. This is usually done by placing one electrode on the dorsum of the therapist's hand and the other electrode on the patient, away from the targeted massage area. Completion of the electrical circuit is accomplished by the contact of the clinician's moistened hand (either hand) on the patient during massage. At the microampere level, there is little or no sensation for either the patient or the physical therapist (Fig. 5-20).

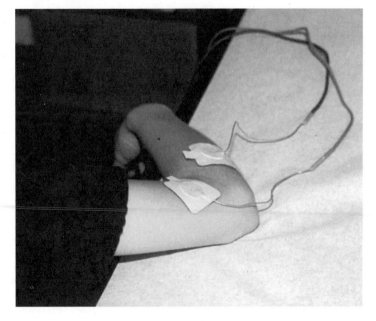

FIG. 5-18. Standard electrodes placed astride the target (epicondylitis) controlled from the generator panel itself. (Courtesy of Monad Corp., Pomona, California.)

Parameters

Parameters must be selected for the individual patient. Treatment goals (i.e., *pain control* or *healing*) must be determined for proper parameter settings. These will include[7]

1. Time
2. Wave slope (ramping ratios)
3. Intensity levels (usually subliminal)
4. Frequency
5. Polarity

Wave slope or ramping is similar in concept to other methods of electrical stimulation. A rapid rise, or sharp wave slope, is usually recommended for normal musculature or acute conditions; a slow rising slope is suggested for chronic

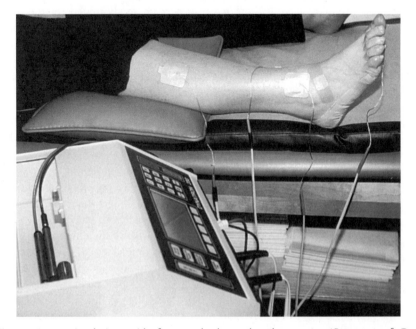

FIG. 5-19. Microampere stimulation with four pad electrode placement. (Courtesy of Dynatron 200, Dynatronics Inc., Salt Lake City, Utah.)

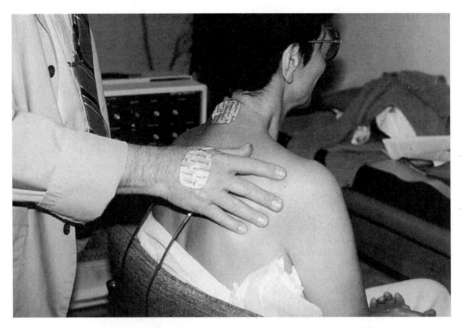

FIG. 5-20. Microampere stimulation, manual technique. Either hand may be used for transmission. However, if *both* hands bear electrodes, *both* hands must be in contact with the patient for continuity of the circuit. (Courtesy of Monad Corp., Pomona, California.)

conditions or regenerating neuromuscular tissues. Examples recommended by Wallace[20] include

1. "Pain" cycle: 30 Hz, 5 minutes, 500 µAmp, slope 2
2. "Healing" cycle: 0.3 Hz, 5 minutes, 100 µAmp, slope 2
3. "Pain/healing": 70 Hz, 5 minutes, 200 µAmp, slope 1 or 2

Summary

It remains to be seen if MES will become another standard modality for physical therapists. Continued research and investigational studies are strongly indicated. Comparison with other means of electrical stimulation is not proper at this point, since the extremely low intensities differentiate MES in concept, physiology, and techniques from traditional methodology. It is an attempt to reproduce the body's electrophysiologic defense mechanism and deserves our attention.

PARAMETERS

A wide selection of parameters is utilized with TENS and EMS (Fig. 5-21). For a comparison of modalities, see Table 5-1.

Frequency or Pulse Rate

High frequency of 80 to 120 Hz (or pulses per second) is recommended for acute conditions when pain is present. At 50 Hz, the normal tetanizing rate, a smooth contraction can be elicited, favorably affecting relaxation of muscle spasm; clinicians have found that a slightly high rate (100 Hz, as mentioned previously) is an optimal frequency for this purpose, since the 50 Hz is a general figure and actually may range from 40 to 120 Hz. The 100 Hz probably will recruit most available fibers in that range.

Low frequency of 1 to 20 Hz is recommended for chronic conditions. Increased endorphin production has been attributed to the lower frequencies; thus, analgesia may accompany the stimulation.[3] (Note: Currents of any frequency may be surged, interrupted, or administered continuously. The likely connection between slow fiber recruitment with surged modes and fast fiber recruitment with interrupted or pulsed modes has been evident to clinicians for some time.)

Pulse Width or Duration

Most equipment available today seems to range from 50 to 400 µs, some ranges being fixed by manufacturers, others decided by clinicians. I

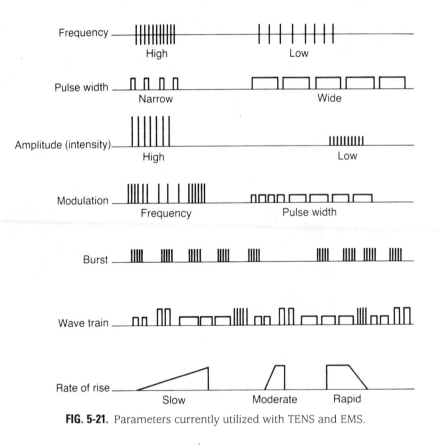

FIG. 5-21. Parameters currently utilized with TENS and EMS.

prefer a medium width of 150 μs, allowing for adjustment in either direction, if indicated. The apparent strength of the stimulating current will seem increased when pulse widths are extended. The increase represents additional energy being provided, however, since the current is "on" for a longer period, rather than there being a direct increase in amplitude.

TABLE 5-1. Comparison of Electrical Stimulation Modalities

Modality	Current	Waveform	Polarity	Purpose
Low voltage	Direct/alternating	Biphasic, monophasic	Alternating, none; direct, ±	Iontophoresis, muscle/nerve stimulation, wound healing
High-voltage pulsed direct current	Direct	Twin-spike, monophasic	Direct, ±	Muscle stimulation, wound healing
Interferential	Alternating	Biphasic	None	Deep muscle stimulation
Russian	Alternating	Biphasic pulsed at 2,500 Hz	None	Muscle stimulation
TENS	Alternating	Biphasic	None	Analgesia, wound healing, non-united fracture
MICROAMP	Direct	Square	± or alternating	Healing, analgesia

Current Amplitude or Intensity

With all electrical stimulation for muscles, I recommend the traditional "visible contraction at patient tolerance" as a clinical guide. The patient's tolerance takes precedence at all times, but a visible (or palpable) contraction of the target musculature should be noted. An exception to this rule is interferential currents, which are often too deep and localized to be seen or felt by the clinician. Patient sensation may be utilized as a guide, however, provided that there are no circumstances that would negate such reports from the patient (e.g., desensitized areas and heavy medication). "The stronger the better" is definitely *not* the rule, because early fatigue and pain from overstimulation could easily lead to additional spasm following the "pain and spasm" cycle, a concept of neuromuscular physiology.

Modulation

Used mainly with long-term applications of TENS for pain control, modulation has not been fully accepted for muscle stimulation. Modulation alters the parameters in order to reduce or minimize accommodation (i.e., the body's adaptation to the stimulating current). Changes in frequency and width, as well as in amplitudes, apparently satisfy the required modifications of current forms. These modulations are most often preset by manufacturers; recently, however, controls have been added to the units, allowing the clinician to make changes as needed in any or all of the parameters. Periodic changes in the frequencies (i.e., 120 to 100 to 120 Hz), in the widths (i.e., 150 to 120 to 150 μs), or a rise and fall in the intensities (approximately 10 percent decrease in preset intensities periodically) fulfill the required modulation procedures.

Burst Phenomenon

Although primarily designed for use with TENS, burst modes have been used for muscle stimulation, since this mode closely resembles the interrupted mode. Small "bursts" or packages of stimuli are administered in a preset intermittent pattern. With intensities adjusted upward to produce visible contractions, this phenomenon may be utilized for muscle stimulation.[21]

Wave Train

Used mainly with TENS, the wave-train form of stimulation is found in some muscle stimulation equipment. Combining and mixing several frequencies, widths, intensities, and modulations of each, this form provides multiple parameters in preset patterns. A useful analogy for understanding this concept is a freight train with many types of cars in a single-linked caravan (i.e., a "train of many waveforms.")

Rate of Rise

The rate of rise is an important parameter with electrical stimulation. It is not as important with TENS. A slowly rising current (i.e., one that reaches peak amplitude over a long time) will usually fail to cause a reaction in normally innervated musculature due to the accommodation factor described earlier. A denervated muscle, however, loses the ability to accommodate and may react to a slowly rising waveform. Often, a faster rise of the current would not be sufficiently long to satisfy the increased requirement of chronaxie widths for muscles exhibiting RD (see Chronaxie section in Ch. 9). Rate of rise, therefore, acquires significance when denervation and regeneration phenomena are involved. In clinical practice, a workable, general rule is to maintain a rapid rise with normal musculature and a slower rise rate with less-than-normal target muscles. Manufacturers usually provide adequate controls for rate of rise alterations, with appropriate scales.

TREATMENT PROCEDURES

Each practitioner develops an individual formula for patient care, consisting of positioning, order of procedures, and modality selection and parameter settings. The following are suggested and recommended techniques and procedures that I have found successful in my practice.

Electrode Placement

Electrode placement is determined by the target muscle or muscle group, either singly or in relation to other muscles and groups. Recruitment is increased when both electrodes are placed on the same muscle. Because this is not always practical, several placement alternatives are recommended below.

Unilateral

Unilateral placement causes stimulation of one extremity or one half of a muscle pair.

Bilateral

Bilateral placement causes stimulation of both extremities or both halves of a muscle pair.

Unipolar

In unipolar placement, the stimulating electrode is placed on target muscle, with the indifferent electrode elsewhere. This technique is usually termed *motor point stimulation* (Fig. 5-22).

Bipolar

In bipolar placement, two electrodes are placed on the target muscle, close to origin and insertion points (Fig. 5-23).

Bilateral Unipolar

Place electrodes on each of two separate muscles or groups (Fig. 5-24).

Reciprocal

Using high-voltage direct current, place an active electrode on each of two separate muscles or groups—either agonist and antagonist or bilaterally—with the indifferent or ground electrode elsewhere (unipolar). Some newer models of high-voltage direct current equipment offer more than the standard three electrodes for multiple target approaches (bipolar technique possible) (see Fig. 5-13).

Interferential

A minimum of four electrodes is required for interferential placement. Place them in a crossed pattern over the target area, approximately 4 to 6 inches (10 to 15 cm) apart. The cruciate pattern may be three-dimensional, that is, medial and lateral and anterior and posterior (e.g., at the knee or shoulder) (Fig. 5-25). Some manufacturers offer vacuum-suction apparatus electrodes for improved skin contacts; however, the advantages of such equipment have not been adequately demonstrated clinically.

Transarthral

Place electrodes on both sides of a joint (e.g., knee, elbow, shoulder, ankle, wrist) in the medial and lateral, the anterior and posterior, or the

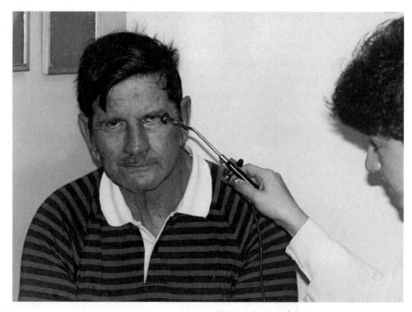

FIG. 5-22. Motor point stimulation (unipolar) with a hand-held interrupter key, using direct current for denervated musculature with a paralysis of the 7th nerve following a brain tumor resection. Finger-interrupted direct current is approximately ¹/₂₅ second in duration, sufficient for most conditions with RD present.

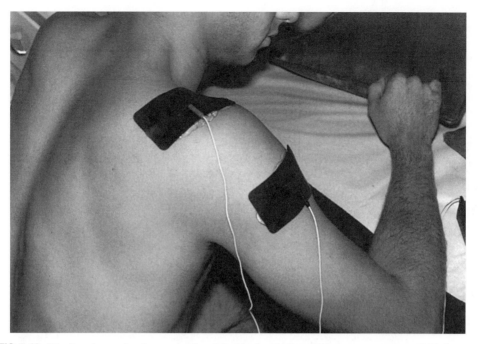

FIG. 5-23. Bipolar electrode placement for deltoid stimulation, with a compact-type EMS device.

dorsal and volar positions. The current does not flow across the joint, as would be expected, but instead flows around the joint between the electrodes. Electric current will flow along the first good conductor (i.e., moist tissues) it encounters, avoiding the high resistances of the intervening structures. In this manner, the entire joint is flooded with stimulating currents, with subsequent recruitment in many muscles. Generally, there are no visible contractions, since there are few muscle fibers in these tendon-rich regions. An advantage, however, is the presence of specific

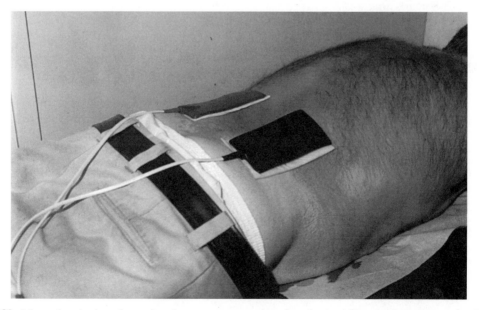

FIG. 5-24. Bilateral unipolar electrode placements over the gluteal, sacroiliac, and quadratus lumborum regions, using a low-voltage unit (sandbags removed). Moistened towels are placed under the rubber electrodes to decrease resistance and improve conformity of the electrodes to body contours. (Intellect VMS, courtesy of Chattanooga Corp., Chattanooga, Tennessee.)

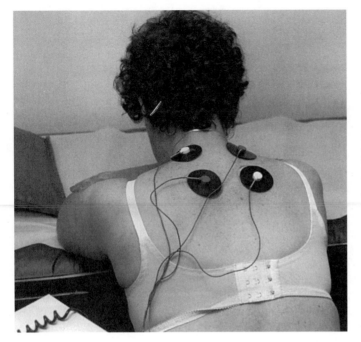

FIG. 5-25. Crossed-pattern electrode placement with interferential current. In this instance of trapezius myositis, the cruciate point is apparently at the C7–T1 level. (Heterodyne, courtesy of Birtcher Corp., El Monte, California.)

structures, Golgi tendon cells, which communicate with the higher centers of the brain along A fibers. These are the *very* fibers targeted for TENS to control pain. Consequently, this transarthral technique often provides analgesia, as well as isometric-type contractions at these joints. This technique is recommended when stimulation *without overt motion* is indicated (e.g., after recent fractures, in cases of severe pain with movement, or immediately postsurgery) (Fig. 5-26).

Special Placements

Placements for special conditions requiring electrical stimulation (e.g., gynecologic, scoliosis) are discussed later in the chapter.[11]

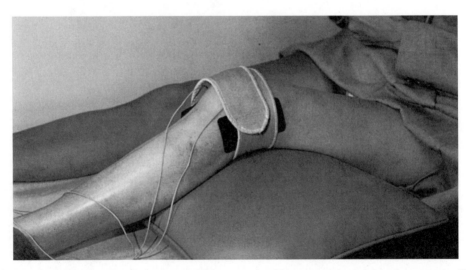

FIG. 5-26. Transarthral electrode placement at the knee. It should be remembered that the current will not run "through" the knee joint. Electricity flows along the first moist layer encountered, so it will flow "around" the knee. In this manner most of the tendons in the area will receive stimulation.

RECOMMENDED TECHNIQUES FOR INNERVATED vs. DENERVATED MUSCULATURE

Whenever possible, bipolar electrode placements are preferred to maximize fiber recruitment in the target muscle(s).

Normally Innervated Musculature

Low-Voltage

1. Tetanize the target muscle with continuous alternating current at 100 Hz for 5 minutes to obtain fatigue and relaxation.
2. Use surged alternating current at 100 Hz, 10 to 15 minutes, 6 to 10 surges per minute.
3. Complete treatment with other modalities as indicated.

High-Voltage Direct Current

1. Tetanize at 120 Hz, with the unit controls on the continuous setting.
2. Follow with 10 to 15 minutes of alternating mode (surged or interrupted) at 5-second alternations.
3. If the patient is athletic, the 10-second alternation mode may be preferable.
4. If the patient is geriatric or debilitated, the 2.5-second alternation may be preferable.
5. Precede or follow this treatment with other modalities as indicated.

Interferential

1. After placing the electrodes in the prescribed crossed pattern over the target area, select the 80- to 100-Hz frequency as a tetanizing mode for 5 minutes.
2. Follow with a "sweep" mode for 10 minutes.
3. Note the patient's report of the strongest sensations at a specific frequency as the sweep traverses the progressive rates. (This may become a diagnostic aid, since recent observations indicate that 60 to 80 Hz is the frequency most commonly reported as the "strong" point with normal musculature, whereas 20 Hz or lower seems to be more prevalent in suspected denervation.)

Denervated Musculature

Low-Voltage

The *only* mode for stimulation to be used with denervation is interrupted direct current (galvanic).

1. Use the hand-held interrupter key, with the polarity set for the stimulating electrode as negative.
2. Place the ground electrode (positive) on the homolateral side of the body to minimize resistance, but not too distant and *not* on an antagonistic muscle group.
3. Stimulate the target muscle at the known motor point, or slightly distally, at 1-second intervals with about 20 repetitions.
4. Rest, then repeat the above procedure several times during each treatment session; patient fatigue, pain, or skin irritation indicates that treatment should be discontinued.
5. Precede or follow with other modalities as indicated.
6. If an automatic, interrupted direct current mode apparatus is available, small standard electrodes may be utilized in place of the interrupter key. The stimulating electrode should always be the smaller of the two pads, even though it is the negative one.

High-Voltage Direct Current

Because the pulses of the high-voltage direct current are too short in duration to be effective with denervated muscles, this modality should not be used for this purpose.

Interferential

Crossed-pattern stimulation with interferential current has not been shown to be effective with denervation since it is not direct current.

Compact EMS Units

Small, battery-operated EMS units, similar in design and size to TENS units, may be utilized when regeneration is manifested clinically. These units are mostly biphasic (alternating current–type) stimulators; all the necessary parameter controls are available to the clinician.

Posttreatment Management

The patient's skin under electrode sites should be cleansed prior to electrode placement. Following stimulation, these areas should be massaged with a bland cream or astringent (e.g., witch hazel, "Köl," carbolated petrolatum) and dusted lightly with talcum powder. If electrical transmission is poor with water-soaked electrodes, a pinch of salt should be added to the water.

RECOMMENDED TECHNIQUES FOR SPECIFIC CONDITIONS

In following the listed treatment regimens, an attempt should be made to obtain visible contractions at patient tolerance whenever possible. Stimulation should not be painful. If visible contractions are painful, subcutaneous, palpable contractions should be obtained instead. These, too, should be painless. All the electrical modalities may be preceded or followed by other indicated modalities during the same treatment session. In addition to the applications listed below, special applications of electrical stimulation (e.g., in cases of temporomandibular joint problems and scoliosis), electrode placement, and recommended parameters can be found elsewhere.[11]

Bell's Palsy

The following regimen is prescribed for Bell's palsy (see Fig. 5-22)[11]:

1. Prior to this treatment, administer iontophoresis with hydrocortisone to decrease the inflammation of the nerve involved.
2. Use interrupted direct current and negative polarity, with the positive electrode on the homolateral forearm.
3. Apply motor point stimulation to the facial (CN VII distribution) musculature, with 10 to 20 stimuli, repeated three times in a session.

Dorsiflexion Paresis

For dorsiflexion paresis with RD, use interrupted direct current and negative polarity, with the positive electrode on the homolateral quadriceps; use 10 to 20 stimuli, rest, and then repeat several times.

Quadriceps Weakness Without RD
Low-Voltage Technique

1. Place bipolar electrodes on the quadriceps (proximal and distal).
2. Tetanize with continuous alternating current at 100 Hz for 5 minutes.
3. Follow with surged alternating current at 100 Hz, six surges per minute, for 10 to 15 minutes (Fig. 5-27).

High-Voltage Direct Current Technique

1. Place both active electrodes (negative polarity) on the anterior thigh, proximal and distal, with the indifferent pad on the lower back (positive).
2. Use continuous mode at 80 to 100 Hz (pulse per second) for 5 minutes.
3. Use alternating mode at 5-second intervals for 10 to 15 minutes.
4. Use surged mode if available and required for slow fiber emphasis.

Interferential Technique

1. Place electrodes in crossed pattern over the belly of the quadriceps.
2. Use continuous mode at 80 to 100 Hz for 5 minutes.
3. Use sweep mode for 10 to 15 minutes.

Compact EMS Units

1. Use same placement as with low-voltage technique.
2. Use tetanizing mode for 5 minutes (continuous). If the continuous mode is not available, set the "on" time to maximum and the "off" time to minimum and run for 5 minutes before changing to the surged or pulsed mode.
3. Use surged or pulsed mode for 10 to 15 minutes.
4. The rate of rise is rapid.

Low Back Spasm
Low-Voltage Technique

1. Place sacroiliac electrode bilaterally; include the quadratus lumborum and proximal gluteal groups (Fig. 5-28).
2. Use tetanizing alternating current at 100 Hz for 5 minutes.
3. Follow with surged mode, six surges per minute, for 10 to 15 minutes.

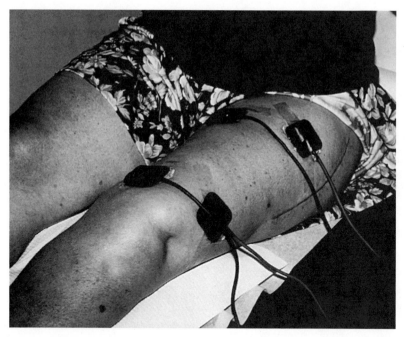

FIG. 5-27. Quadriceps stimulation using four electrodes for maximum recruitment. (From Kahn,[11] with permission.)

High-Voltage Direct Current Technique

1. Place both active electrodes over sacroiliac regions (negative).
2. Place positive ground electrode on the dorsal spine.
3. Tetanize at 80 to 100 Hz for 5 minutes.

4. Follow with alternating mode at 5 seconds for 10 to 15 minutes.

Interferential Technique

1. Place electrode in crossed pattern over lumbo-sacral region, approximately 6 to 8 inches (15

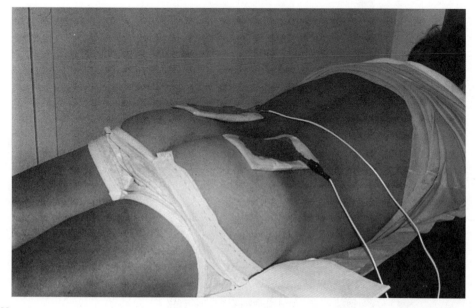

FIG. 5-28. Bilateral, low back electrode placements, similar to that described in Fig. 5-19, except that paper towel and aluminum foil electrodes are utilized and sandbags are removed. Tetanizing alternating current to relax the spasmodic musculature precedes surging alternating current.

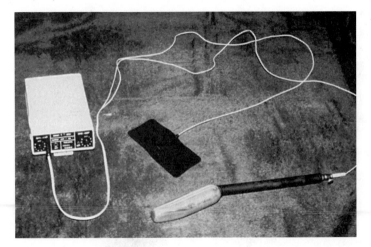

FIG. 5-29. A vaginal electrode is connected to a compact EMS unit for use at home. (Courtesy of Mentor Corp., Minneapolis, Minnesota.)

to 20 cm) square, with the cruciate point over the lumbosacral joint.

2. Use continuous mode at 80 to 100 Hz for 5 minutes.

3. Use sweep mode for 10 to 15 minutes.

Compact EMS Units

1. Use same placement as for interferential technique for the four electrodes if a dual-channel electrical muscle stimulation device is used or bilateral proximal gluteal area if a single-channel (only two electrodes) unit is used (Fig. 5-29).

2. Use continuous or semicontinuous mode, 5 minutes (see Chemical Effects).

3. Use intermittent mode for 10 to 15 minutes.

Stress Incontinence (Vaginal Stimulation)[11,22–25]

1. Prewarm the intravaginal electrode, coat it with transmission gel, and slowly insert it into the vagina (Fig. 5-30).

2. Place the indifferent electrode in the suprapubic region and secure it with a lightweight sandbag.

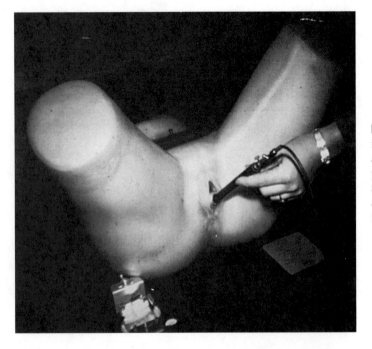

FIG. 5-30. A warmed, lubricated vaginal electrode is inserted appropriately, with a dispersal electrode placed in the suprapubic region. Surged alternating current is preferable in the management of stress incontinence.

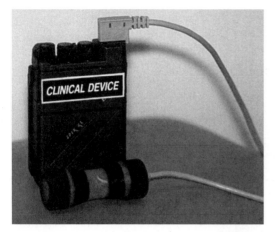

FIG. 5-31. Vaginal electrode connected to EMPI's EPIX unit. The two outer bands are one circuit; the two inner bands are the second circuit. Conductive gel lubrication facilitates insertion.

3. Administer surged alternating current at 100 Hz, six surges per minute, for 10 to 15 minutes, twice weekly.
4. Follow with 10 minutes of shortwave diathermy (if available) through the anterior abdomen. Teach and stress the importance of pelvic floor exercises.

A new type of vaginal electrode has been introduced by the EMPI Co. It is used in conjunction with their electrical stimulation unit, EPIX, or with a specially designed "patient-friendly" model for home use (Fig . 5-31).[26,27]

Scoliosis

Although specialized units are now available for the stimulation of musculature involved with idiopathic scoliosis, physical therapists have been using electrical stimulation for this condition for many years (Fig. 5-32).[11,28–30] Electrode placement has varied from paravertebral to far lateral placements, with equally variable results. I have found the following procedure the most effective:

1. Place electrodes paravertebrally, approximately 2 inches lateral to the spine, with both electrodes on the *convexity* of the curve, at the proximal and distal portions.
2. If the curve is a compensated *(S)* curve, use four electrodes, with two on each convexity.
3. Administer surged alternating current with a frequency of 80 to 100 Hz.

4. Set pulse width at 150 μs.
5. Rise is rapid.
6. Set controls at *on* for 3 seconds and at *off* for 3 seconds.
7. Intensity is minimal, but produces visible contractions of the spinal and posterior chest wall musculature.
8. Instruct the patient to administer the stimulation nightly for 8 hours, while sleeping.
9. Utilize stimulation instead of or in addition to standard bracing.
10. Follow with radiologic checkup and reevaluations at 3-month intervals to monitor progress.

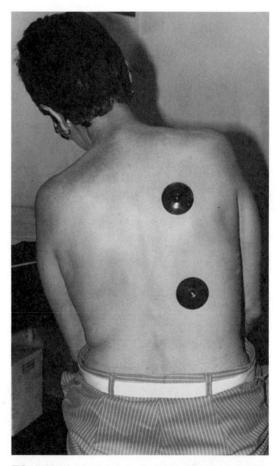

FIG. 5-32. In the electrical treatment of idiopathic scoliosis, electrodes are placed on the convexity of the curve (or curves). Far lateral placements are also recommended; promising results have been reported. Clinicians should consider either or both, with regard to their provision of comfort and their effectiveness. Parameters and prescription data are discussed in the text. Compact-type EMS units are indicated for idiopathic scoliosis.

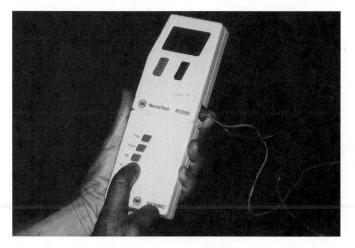

FIG. 5-33. Computerized electrical stimulation device. The program is entered, the computer is disconnected from the stimulator, and the patient takes the stimulator home for use. (Courtesy of Neurotech 2000, BMR, Phoenix, Arizona.)

RECENT DEVELOPMENTS

A relatively recent addition to the electrical stimulation armamentarium is a computerized, programmable unit for home use. A miniature computer is connected to the stimulator (Fig. 5-33). The clinician programs desired parameters into the device, which is then disconnected from the computer and taken home by the patient. A wide variety of parameters is available, with several preselected protocols recommended by the manufacturer.[31] It should be noted that although protocols are "standardized," our patients are not; clinicians should adjust, change, and modify these programs for each patient.

There is still a need for pulsed direct current circuits for treatment of denervated musculature. At present few units are available.[32]

Management of open wounds with high-voltage electrical stimulation is being recommended by many clinicians. The therapist should search current literature for pertinent references.[33]

Incontinence following a radical prostatectomy may be addressed with electrical stimulation via the rectum. Traditional electrodes for this purpose are still available.[34] Reference electrodes are placed on the gluteal or quadriceps muscles (Fig. 5-34). Innovative specialized devices are also available (Fig. 5-35). For stress incontinence in women, vaginal electrodes are advised, as described earlier in this chapter (see Figs. 5-29 and 5-30).

I recently investigated the efficacy of microampere stimulation for recession of the gums (my own). Two probe electrodes were utilized, allowing the current to pass between them along the gum line. Alternating polarity was selected for the advantages of both the positive and negative fields. The goal was to increase the circulation to the shrinking gums (Fig. 5-36). Results were favorable.

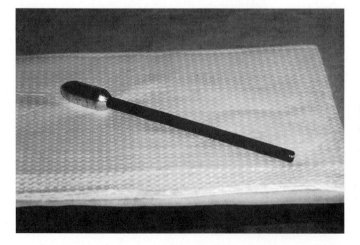

FIG. 5-34. Traditional vaginal electrode, adapted for a pin-lead stimulator.

Another innovation is the "glove" and "stocking" electrode system (Fig. 5-37). This method is valuable for treating generalized pain and edema in the extremities, since the stimulation is both dorsal and ventral. When only dorsiflexion or volar flexion is desired, direct bipolar stimulation techniques are advised.

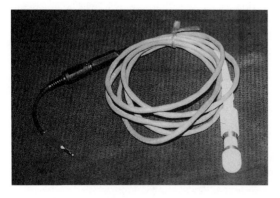

FIG. 5-35. Special rectal electrode.

FIG. 5-36. Microampere stimulation with probes for gum recession. Current flows between probes along the gum line, increasing capillary flow.

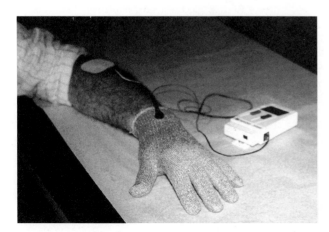

FIG. 5-37. "Glove" electrode for general hand stimulation. A "sock" is also available.

REFERENCES

1. Kahn J: The Basics of Electrical Stimulation. Whirlpool, Fall 1984, p. 26
2. Shriber WJ: A Manual of Electrotherapy. 4th Ed. Lea & Febiger, Philadelphia, 1981
3. Olsen DW: Low Frequency Stimulation of Acupuncture and Trigger Points for the Relief of Pain. Empire State Physical Therapy. NY Chapter/APTA. November 1982, p. 10
4. Smith CW, Aarholt E: Effects of endogenous opiates from electricity. Commentary by Kohn II. Chemical & Engineering News, February 14, 1983, p. 59
5. O'Brien WJ, Rutan FM, Sanborn C, Omer GE: Effect of TENS on human blood beta-endorphin levels. Phys Ther APTA 64:1367, 1984
6. Owens J, Malone T: Treatment parameters of high frequency electrical stimulation as established on the electro-stim 180. JOSPT, Winter 1983, p. 162
7. Alon G: High voltage stimulation—effects of electrode size on basic excitatory responses. Phys Ther APTA, 65:890, 1985
8. Kahn J, Schell F: Clinic's electrode technic dries up hyperhidrosis. Medical Group News, September 1973, p. 24
9. Harvey JS: Basic Muscle Physiology. Fort Collins Sports Medicine Clinic, Ft. Collins, CO (undated)
10. Kloth L, Feedar JA: Acceleration of wound healing with high voltage monophasic pulsed current. Phys Ther 68:503, 1988
11. Kahn J: Clinical Electrotherapy. 4th Ed. J Kahn, Syosset, NY, 1985, p. 43
12. Goulet MJ: Interferential current now being introduced as a modality. p. 1. Physical Therapy Forum, Vol. 111; No. 37. September 12, 1984, King of Prussia, PA
13. Diadynamics: B.V. Enrauf-Nonius, Delft, Holland, March 1985
14. Nordenstrom B: Biologically Closed Electric Circuits: Clinical, Experimental and Theoretical Evidence for an Additional Circulatory System. Nordic Medical Publishers, Stockholm, Sweden, 1983
15. Picker RI: Current trends. Clin Management 9:10, 9:28, 1989
16. Gersh MR: Microcurrent electric stimulation: putting it in perspective. Clin Management 9:51, 1989
17. Gault W, Gatens P: Use of low intensity direct current in management of ischemic skin ulcers. Phys Ther 56:265, 1976
18. Wallace L: MENS Therapy. 4 Vols. Ohio Biomedical Education Research, Cleveland, OH, 1990
19. Becker RO: The Body Electric. William Morrow Co., New York, 1985
20. Wallace L: New Perspectives in Rehabilitation and Preventative Health Care Services. MENS Therapy. Lyndhurst, OH, 1986, p. 20
21. Brill MM, Whiffen JR: Application of 24-hour burst in a back school. Phys Ther APTA, 65:1355, 1985
22. Malverne J, Edwards L: Electrical stimulation for incontinence. Am Urol Assoc, abstracted in Medical Tribune, June 20, 1973, p. 24
23. Moogaiker AS: Management of stress incontinence in women. Geriatrics, June 1976
24. Godec C, Cass A: Acute electrical simulation for urinary incontinence. Urology 12:340, 1978
25. Hudgins AP: Electrical muscle stimulation in gynecology. Am Pract 6:1695, 1955
26. Kahn J: Electrical stimulation used as treatment for stress incontinence. Phys Ther Adv 3:10, 1992
27. Alon G, Fredrickson R, Gallagher L et al: Electrical stimulation of the abdominals. The effect of 3 vs 5 weekly treatments. J Clin Electrophysiol 4:5, 1992
28. Axelgaard J, Norwall A, Brown JC: Correction of spinal curvatures by TENS. Spine 8:463, 1983
29. Eckerson L, Axelgaard J: Lateral electrical surface stimulation as an alternative to bracing in the treatment of ideopathic scoliosis: treatment protocols and patient acceptance. JAPTA 64:483, 1984
30. Surface muscle stimulation for spinal deformities; an alternative to bracing. Rancho Los Amigos Hospital Engineering Center, AAOS, Downey, CA, 1982 exhibit
31. BMR Neurotech Co, Phoenix, AZ
32. Staodyne Corp, Longmont, Co. "EMS;p12."
33. Luther Kloth PT, Marquette University, Milwaukee, WI.
34. Electro-Med Health Industries Inc, Miami, Florida

SUGGESTED READINGS

APTA: Electrical Stimulation Anthology. Vol. 4 APTA, Alexandria, VA, 1993

APTA: Electrotherapeutic Terminology in Physical Therapy. APTA, Alexandria, VA, 1990

Benton L, Baker LL, Bowman BR et al: Functional Electrical Stimulation. Rancho Los Amigos Hospital, Downey, CA 1980

Forster A, Pastalanga N: Clayton's Electrotherapy. 8th Ed. Ballière Tindall, London, 1981

Gersh MR: Electrotherapy in Rehabilitation. FA Davis, Philadelphia, 1992

Low J, Reed A: Electrotherapy Explained. Butterworth-Heineman, London, 1990

Kahn J: Ampere Ambience, Empire State Physical Therapy, April 1996, p. 8

Kahn J: Electric Stimulation Used as Treatment for Stress Incontinence. Phys Ther Adv, March 9, 1992, p. 5

Kahn J: Electrical Stimulation (videotapes). Neurotech Corp., 85 Flagship Dr., Suite A, North Andover, MA 08145

Kahn J: Electrotherapy with Bell's Palsy. Platform Presentation at APTA CSM. Dallas, TX, February 15, 1997

Kahn J: Post Prostatectomy Incontinence: Electric Stimulation. Platform Presentation at APTA CSM. Dallas, TX, February 15, 1997

Kaplan PG, Tanner E: Musculoskeletal Pain and Disability. Appleton & Lange, East Norwalk, CT, 1989

Scully M, Barnes MR: Physical Therapy. JB Lippincott, Philadelphia, 1989

Taubs G: An electrifying possibility. Discovery, April 1986, p. 22

Wadsworth H, Chanmugan A: Electrophysical Agents in Physiotherapy. Science Press, Marrickville, Australia, 1980

Wilder E: Obstetric and Gynecologic Physical Therapy. Churchill Livingstone, New York, 1988

6

Transcutaneous Electrical Nerve Stimulation

Introduced to the profession in the early 1970s, transcutaneous electrical nerve stimulation (TENS) has rapidly been accepted as a standard modality in the management of pain, both chronic and acute. Electrical stimulation controls pain noninvasively and without narcotics. TENS, therefore, is a specialized form of electrical stimulation that is designed to reduce pain, in contrast to other forms of electrical stimulation used either to produce muscle contractions or to introduce chemicals into the body (iontophoresis). The compact, portable TENS devices available today are ideal for patient use at home, offering continuity as well as continuation of care once the patient is out of the hospital. Initial reports comparing TENS with narcotics to manage pain indicated that the new method was a viable alternative to narcotics, provided that the patient was not already addicted or heavily dependent on drugs for pain relief.[1] Because of its wider range of clinical applications, TENS quickly became known to many in the medical and allied health professions. Podiatrists, dentists, and veterinarians soon discovered the benefits of applying TENS to their clinical needs. Physicians, in general, have been reluctant to accept TENS as a standard procedure and most often resort to this modality after all else has failed. This "last resort" approach naturally minimizes effectiveness, particularly if the patient is already heavily dependent on medication and the condition itself has advanced to a difficult stage. Those physicians (and others) who have utilized TENS properly have reported excellent results in many aspects of practice: presurgical and postsurgical pain,[2] nonunited fracture pain and healing,[3] obstetrics,[4,5] dental and temporomandibular joint pain,[6] and a multitude of podiatric pain applications.[7] Although there have been anecdotal reports, professional literature regarding applications in veterinary practice is still lacking.

SELECTING A TENS UNIT

Since the TENS unit first appeared in the health care industry, the field has grown to include more than 50 manufacturers, with multiple brands and models from which clinicians may select their favorite unit (Fig. 6-1). Each model is slightly different, offering various parameters, equipment, sizes, shapes, and commercial advantages. Most clinicians agree that TENS, when applied properly, is effective, regardless of the brand or model. Conversely, in untrained hands, most TENS units are not an effective or valuable modality. TENS is a tool that must be utilized in a logical manner to produce favorable results.

I have not found that any one brand or model is sufficiently superior to the others to warrant its exclusive use in my facility. I have found, however, that one or two units are generally more effective and acceptable for my practice, and I tend to utilize, prescribe, and recommend them more often than others. Clinicians should be able to evaluate and decide which brand, model, and unit they think is best. I suggest, however, that this evaluation be made based on clinical evidence rather than on sales or marketing claims. A manufacturer who produces a reliable, effective, and reasonably priced product usually does not fear competition through comparison. Fortunately, in this area, there is room for all, since the tastes and personal idiosyncrasies of physical therapists are so diverse. It is the results with patients that count and make the difference.

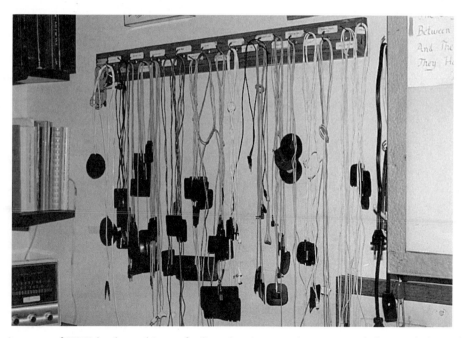

FIG. 6-1. An array of TENS leads used in our facility, showing round, square, and elongated electrode types.

PHYSICS AND PHYSIOLOGY

The Gate Theory

With TENS, an electrical current is applied to nerve endings in the skin, which travel toward the brain along selective nerve fibers (i.e., A fibers) or proprioceptive spacial-location data gatherers. According to the pain theory of Melzack and Wall,[8] these fibers must pass through a segment of the spinal cord, the substantia gelantinosa, which contains specialized cells involved in neu-ral transmission. These T cells also serve as transmission junctions for those nerve fibers carrying the sensation of pain upward toward the thalamus, or "pain center" of the brain (Fig. 6-2). The small c fibers offer a transmission velocity that is considerably slower than that of the A fibers. Thus, the signal along A fibers normally reaches the brain before the c transmission. Both fibers and their respective transmissions must pass through the same T cells in the spinal cord, as mentioned, with a preponderance of A fiber input

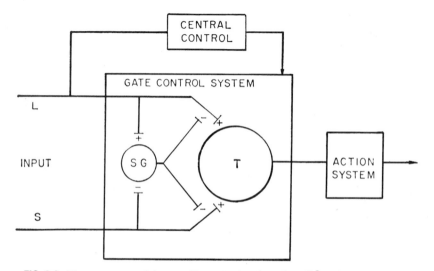

FIG. 6-2. The gate control theory. (From Melzack and Wall,[8] with permission.)

due to the great numbers present in the system and the rapid rate of transmission. If the T cell is considered as a gate through which these signals must pass, it is conceivable that an overload of A transmission could block the incoming, slower moving c transmission carrying the pain signal to the brain. In this manner, a pain signal could be effectively blocked by the gating mechanism described within the T cell. Pain would therefore be decreased or blocked entirely for the patient. This is the basic concept of the gate theory of Wall and Melzack.

Production of the Gating Effect

To produce the gating effect well, we must increase A fiber transmission *without* increasing c fiber (pain) transmission. How this is accomplished is a tribute to engineering acuity and industrial ingenuity. Researchers found that A fibers respond, to a greater extent than do other fibers, to phasic input (i.e., waveforms that are not sensed by the body as continuous and generally contain multiple phases of positive and negative modes). In contrast, c fibers apparently react best to continuous waveforms or to those sensed by the body as continuous. For example, a high-frequency alternating current may be too high in frequency for the body to distinguish individual stimuli and so may be felt as a continuous form. The accepted threshold or tetanizing frequency for normal human systems is approximately 30 to 50 Hz.

Consequently, clinicians were able to stimulate the surface of the skin with a phasic input device to stimulate those select A fibers without disturbing the c fibers, which led to the development of the TENS modality. Ramifications and variations in the parameters of the devices followed later and blossomed into the huge competitive TENS industry of today.

Endorphin Concept

I do not attempt in this book to discuss the details of neurophysiology of endorphin chemistry. The professional literature makes sufficient reference to that aspect of pain modification.[9-11]

The body produces endorphin, a morphinelike molecule, to serve as an endogenous analgesic whenever the body senses pain. Blood levels of this substance increase when the incoming signals to the brain indicate the presence of pain.

TENS research indicates that endorphin production may be enhanced by electrical stimulation, producing a painlike reaction effect on the cells producing the endorphin. The stimulation does *not* have to be painful to produce this effect. In previous sections, I alluded to a differential reaction to low- and high-frequency stimulation with regard to the endorphin concept. Yet to be proven in the laboratory, but obvious in the clinic, is the greater effectiveness of low rather than high frequencies in increasing endorphin production. Therefore, low frequencies (e.g., 1 to 4 Hz) are suggested in chronic conditions in patients whose available supplies of endorphin have been depleted over months and perhaps years of pain.[12] In contrast to the "brief/intense" concept mentioned earlier, the extended duration of stimulation offered by traditional TENS techniques capitalizes on the increased endorphin output for prolonged relief.

Very low rates (e.g., 1 or 2 Hz), when administered in increasing intensities, may be utilized for muscle stimulation, relying on the endorphin production to act as analgesia. Very few TENS units, however, offer frequencies that are low.*

PRECAUTIONS

1. Do not place electrodes in the area of the carotid artery (sinus) in the anterolateral region of the neck.
2. Do not use TENS on a patient with a demand-type pacemaker in place.[13]
3. Do not administer TENS for undiagnosed pain. Pain is an important diagnostic symptom that should not be masked.
4. Athletes should not be permitted to participate in sports while under the influence of TENS analgesia. (This rule is applicable in veterinary practice with regard to animal competitions.)
5. Extreme caution is needed with patients under the influence of narcotic medication or who are known to have hyposensitive areas.
6. Caution is recommended in using TENS for pain control for a pregnant patient other than for labor and delivery.

---

*PMS-5 TNS unit, Wallant International Trade Corp., 312 Cox Street, Roselle, NJ 07203.

PARAMETERS

Waveforms

Biphasic

Current TENS models favor the biphasic waveform, containing both the positive and negative phases. These waveforms can be square, rectangular sinewave, or triangular/spiked (see Fig. 5-12). In most instances, efforts are made to equalize the positive/negative phases to maintain either a net direct current component of zero or no electrochemical effect due to excessive polarity. It has been shown, however, that the difference in intensity of one of the phases—usually the positive—offers a slight bias in that phase.[14] Although this bias is not enough to be of clinical significance, it does necessitate the color coding of the electrode leads (e.g., red [+] and black [−]). To minimize the direct current effect, manufacturers have produced units with a compensated rectangular biphasic wave as well as units with compensated asymmetrical spike waves. When a unit contains red and black electrode leads or the operational manual recommends reversal of electrode placements for increased effectiveness, the clinician can safely assume a polarity differential between electrodes, regardless of claims of compensation and a net direct current = 0 component.

Waveform Selection

To date, there has been little or no clear evidence of physiologic benefit of any specific waveform other than some ability to provide patient comfort.[15] In some instances, however, the pathology demands a rapidly rising form (e.g., the spike wave), whereas others require a longer duration (rectangular or square form). Spike waves are generally more irritating to the skin and often require frequent movement of electrodes or shorter treatment times to avoid skin irritations.

For hypersensitive patients, the square or rectangular wave is recommended. These longer duration waveforms are also suggested when some nerve damage has been associated with the pain pathology. As these waveforms approach the shape of the sinewave, the skin irritation is less.

Spike waves are recommended for intense or hyperirritating stimulation, such as would be administered for acute pain or resistant tissues.[16,17] I have found that intense stimulation with spike waves does not produce as long-lasting a relief as that provided by the longer duration square or rectangular types. Perhaps clinicians should use the sharp, spike wave for immediate, temporary relief with acute pain and the longer duration square, rectangular, or sinewave forms for chronic pain patients to provide delayed and longer-lasting analgesia.

Frequency or Rate

Frequency or rate controls indicate the number of stimuli being transmitted each second. Generally, the frequencies are considered high when TENS is used—in the range of 80 to 120 Hz. Such frequencies are selected if the condition is acute, and they tend to offer more immediate relief. The lower rates, in the range of 1 to 20 Hz, are more applicable in cases of chronic pain.[18] The possible explanation for this difference is discussed in the section on the endorphin concept of pain modification.

Pulse Width or Duration

The extent of the waveform in time—that is, the length of the time the current is actually acting on the patient in each individual pulse—is measured in microseconds and, in current models, ranges between 50 and 400 µs. When TENS is applied to a normal neuromuscular system, a general range of 100 to 150 µs is recommended, whereas in patients with neurologic damage, wider widths are indicated (e.g., 200 to 300 µs). If neurologic damage with reaction of degeneration (RD) is present, biphasic waveforms will not normally elicit muscle contractions, and the clinician must rely on monophasic, interrupted direct current to treat these conditions. When electrical stimulation is administered *not* for muscular contraction, but for pain modification, the biphasic waveform may be utilized, however. Increased durations are recommended because of the less-than-normal status of the damaged nerve (see Chronaxie section in Ch. 8).

Amplitude or Intensity

Low Amplitude

Controversy still exists among clinicians as to the "ideal" intensity for TENS administration.[16] I prefer an intensity that is barely sensed by the patient—low enough so that if it were to fade out entirely the patient would immediately sense the loss of stimulation. I have found that increased intensities mask any fading out and leave the patient unaware of decreased stimulation, usually due to the body's accommodation. The clinician should be alert to the fact that dying or dead batteries can be the cause of fading intensities.

Brief-Intense Amplitudes

A recent preference for the brief-intense technique has added yet another approach to treatment with TENS.[15] In this case, the high-amplitude/narrow-width combination offers more immediate relief than do other forms. My own clinical experience, however, has shown that the relief afforded by use of this method is often too short-lived compared with the longer-lasting benefits offered by the lower intensities (see Endorphins section).

Ranges

Most TENS units range from 1 to approximately 100 mA. Treatment should be based on sensation, however, rather than on milliampere readout. Because skin resistance remains a relative factor, especially with the transmission gel's effect of decreasing resistance, the relationship between the current and the voltages follows exactly the traditional dictates of Ohm's law: $I = E/R$ (current equals voltage divided by resistance).

Constant Current vs. Constant Voltage

Some manufacturers offer a "constant current" with their products, which, in effect, means that the current will remain constant despite variations in resistance by consequent automatic adjustments in the operating voltage. Other manufacturers state that their units operate at "constant voltage," which means that the voltage will remain constant despite the same resistance changes, with a concurrent automatic adjustment of the milliamperage. The obvious advantages and disadvantages of each system should be evaluated clinically by the practitioner. For example, a patient's twisting in bed may disrupt the skin contacts of the electrodes and increase the resistance in the operating circuit. With a constant-current model, the internal voltage will automatically increase to account for the higher resistance to maintain the selected amperage. The increased voltage offers continuity to the treatment, but may increase skin irritation. On the other hand, if a constant-voltage unit is used in the same situation, the milliamperage will drop with the increased resistance, offering current levels that have little efficacy but that do cause skin irritation. Most modern units are constant current.

Modulation

One should remember that the body will eventually get used to anything. Electrical stimulation can become less effective as the body accommodates itself to the current and reacts less to its stimuli. Slight alterations in the course of treatment can overcome this accommodation tendency. Changes in any of the parameters may be employed to vary the current flow for this purpose (see Fig. 5-16).

Modulating the Treatment

Operating manuals for each model generally offer complete descriptions of the parameters and modulations available.

Frequency Modulation

Intermittent changes in the frequency, as set by the manufacturer or the clinician, are usually programmed to vary about 10 percent periodically (e.g., 100 to 90 to 100 to 90 Hz, etc.).

Pulse Width Modulation

Intermittent changes in the pulse width, as set by the manufacturer or clinician, are usually programmed to vary about 10 percent periodically (e.g., 150 to 135 to 150 to 135 μs, etc.).

Amplitude Modulation

Intermittent changes in the amplitude, as preselected by the clinician, also vary about 10 percent (e.g., 10 to 9 to 10 mA, etc.). This is not a popular technique; because the treatment is usually based on patient comfort and sensation, alterations in the amplitude may lead to discomfort.

BURST MODULATION

The burst phenomenon provides a packaging of several stimuli in groups ranging from 1 to 10 and is presented in "bursts" of energy sensed by the patient as a single stimulus (see Fig. 5-16). Whether this mode offers any physiologic advantages has not yet been evaluated.

Wave Train

Some manufacturers have designed a series of modulations, bursts, and standard modes in a preprogrammed (controlled or fixed) flow, in the form of a *wave train*. This series of modulations is analogous to a freight train that has repetitions of different types of cars in a prescribed order. The physiologic benefits of this mode have not been established, although the following may be said: all modifications and modulations tend to increase the "phasic" quality of the current flow; thus, they tend to increase the reception by the

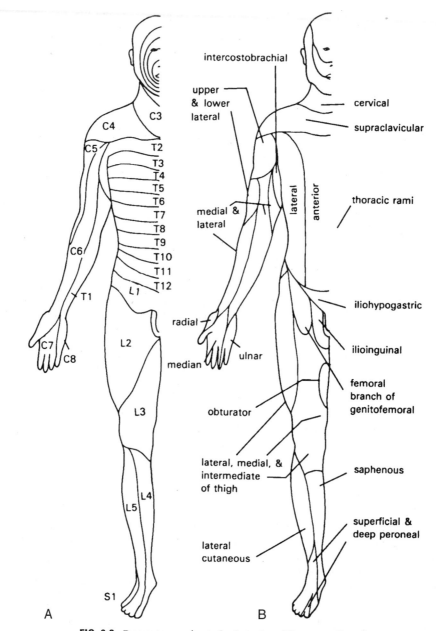

FIG. 6-3. Dermatome chart. **A,** Anterior. *(Figure continues)*

A fibers, thereby gradually enhancing the pair modification effects.

TREATMENT PROCEDURES

It has been my policy to prescribe TENS for pain control four times daily, 1 hour per session. As pain diminishes, sessions may be decreased to three times, twice, and once daily, as needed. Occasionally, I permit constant use when a patient is traveling a long distance by plane or automobile. I generally prefer, however, to allow the skin and other related systems a rest period without stimulation. Nonetheless, for postoperative use, painful scars, and obstetrics, prolonged

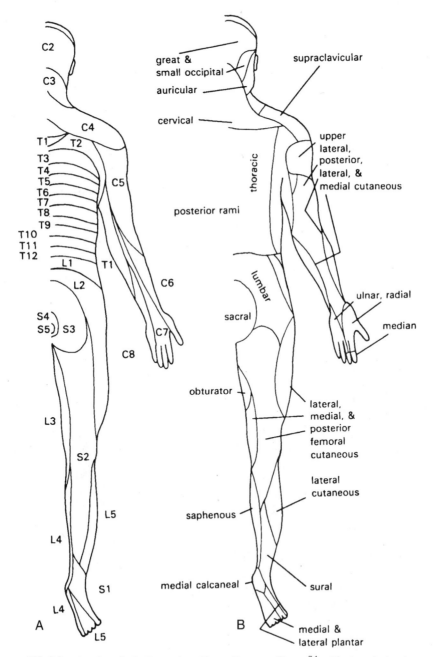

FIG. 6-3. *(Continued).* **B,** Posterior. (From Romero-Sierra,[34] with permission.)

use is advised; when used for nonunited fracture healing (see Obstetrics and Nonunited Fractures sections) the 1 hour, four-times-daily prescription is recommended.

Treating Acute and Chronic Pain

Prior Medication

If the patient has been on heavy medication prior to starting TENS, sufficient time should be allowed for the effects of the drugs to diminish, usually a week or so, before the effectiveness of the TENS program is evaluated.

Prepping the Patient

Skin in the area of electrode placements should be clean and clear of lesions. Standard electrical conduction gels or sprays are recommended, except with those TENS units designed to use water as a conduction medium, with sponge electrodes. Commercial tapes or household mending tape* may be used to secure electrodes in position for short-term use. Longer applications may necessitate special tapes or electrodes (e.g., karaya or other conductive materials) to maintain contact throughout the long period of stimulation, perhaps for hours or days. These products are usually available from manufacturers of the TENS unit.

Electrode Placements

Probably one of the most controversial topics with TENS is the question of ideal electrode placement. I do not believe there is any exact site for electrode placement. Many techniques are suggested, based on nerve roots, acupuncture points, and trigger points (Fig. 6-3); all are valuable but vary with each individual case. Unless the clinician is unusually gifted, experimentation is suggested so that the optimal position for the electrodes may be found in each individual case. The use of electrical probes is sometimes effective in locating tender or key points. An experienced practitioner with TENS will quickly be able to establish several key anatomic points to cover most conditions. The following are some recommended electrode placements for common conditions.

---

*Mending tape, 3M Corp., Minneapolis, MN.

Placement of Electrodes for the Upper Extremity (Fig. 6-4)

1. C3–C7 nerve roots/dermatomes
2. Point of pain
3. Tip of the acromion
4. Hoku (web space between thumb and forefinger)
5. "Wristwatch" position, dorsal wrist
6. Tip of lateral epicondyle

Placement of Electrodes for the Lower Extremity (Fig. 6-5)

1. L1–S1 nerve roots/dermatomes
2. Gluteus maximus center ("bull's eye")
3. Popliteal space
4. Posterior lateral malleolus
5. Head of the fibula
6. Specifically for the knee: transarthral medial and lateral aspects

Placement of Electrodes for the Lower Back (Fig. 6-6)

1. Associated nerve roots and dermatomes
2. Gluteal sites as above
3. Popliteal sites as above
4. Crossed pattern: paravertebral at L1 and L5, in a boxlike pattern, with the circuits crossing at L3

General Configurations (Fig. 6-7)

1. Associated nerve roots and dermatomes
2. Point of pain
3. Acupuncture point proximal to point of pain
4. Acupuncture point distal to point of pain
5. If pain can be pinpointed, consider the *crossed-pattern* technique above with the crossing point at the painful site.
6. Transarthral placements are effective at the shoulder, knee, elbow, wrist, and foot.
7. Bilateral placements are extremely effective when practical, especially with midback and low-back conditions.
8. Contralateral placements are suggested when the pain site is not accessible due to amputation, dressings, open wounds, and casts.
9. Rarely are more than four electrodes needed. If the clinician cannot produce analgesia with four electrodes, more are unlikely to be of value.

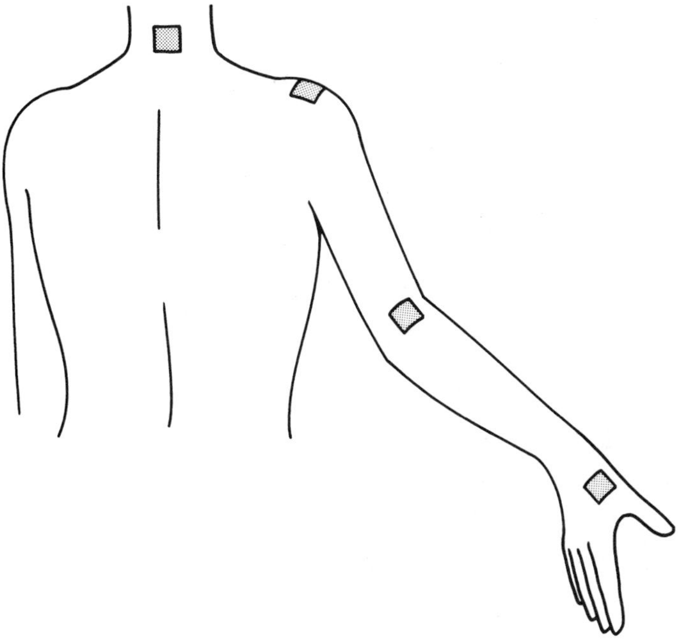

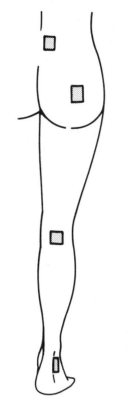

FIG. 6-4. Upper extremity electrode placements: nerve root, acromion tip, lateral epicondyle, and hoku.

FIG. 6-5. Lower extremity electrode placements: nerve root, gluteal, popliteal, and posterior lateral malleolar.

RECOMMENDED TECHNIQUES FOR SPECIFIC CONDITIONS

Nonunited Fractures

A nonunion is usually described as a fracture of 6 months' duration with no healing.

Electricity and Osteogenesis

Electricity has long been known to be a stimulus for osteogenesis. The literature of our profession contains many references to earlier research. The current surge in our field was felt in the 1970s. Bassett and colleagues[19] worked with a variety of current forms to enhance union of fractures. These included high-frequency, pulsed direct current and continuous modes of alternating and direct current. Success was moderate, and favorable reports are found in the literature.[19,20] Other physicians, using surgical intervention, applied current directly to the fracture sites through needlelike electrodes placed internally into the area of the nonunion. All researchers reported favorable results.[21]

TENS and Osteogenesis

In 1980, a fortunate occurrence led to my applying TENS for the same purpose. A young woman with a nonunited tibial fracture of 6 months' duration was referred to me for TENS as a

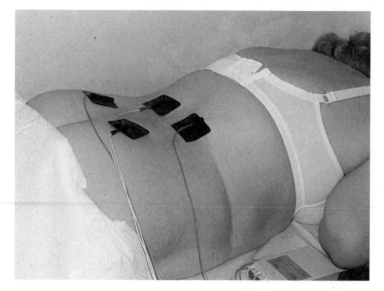

FIG. 6-6. Lower back electrode placements (no radiculopathy): crossed-pattern with gluteal and nerve root sites; crossing point at S1. (Wallant PMS-10, Roselle, New Jersey.)

possible pain control modality prior to planned surgical intervention. A fortunate delay in surgical scheduling and favorable results in pain control with TENS allowed an entire month of treatment. At the end of the month, a routine x-ray film was taken to check on the status of the nonunion. We were quite surprised to find almost complete bridging with callous and early reossification at the fracture site.[22] Surgery was cancelled, and the TENS unit was continued until complete ossification had taken place. With additional nonunion cases, similar results were achieved. With the exception of those cases involving osteotomies or infections, most clinicians report success with TENS within 4 to 6 weeks. Osteotomies and infections require much more time to show favorable changes, if any. All case results are confirmed by radiologic evidence.

TENS' Place in Nonunited Fracture Treatment

It may be proven that the actual current mode (high or low frequency), pulsed alternating or direct current, or waveform makes little difference. The electrical energy and the bone itself may be the key factors in determining the outcome of nonunited fractures. Nevertheless, TENS is a successful modality in bone healing; the research details must follow. It is known that bone exhibits piezoelectrical qualities—similar to those of natural or manufactured crystals—and that currents are generated, albeit in the low

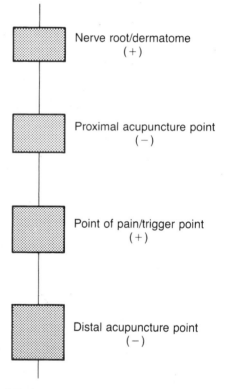

Nerve root/dermatome
(+)

Proximal acupuncture point
(−)

Point of pain/trigger point
(+)

Distal acupuncture point
(−)

FIG. 6-7. General formula for electrode placements: nerve root, proximal acupuncture point, point of pain or trigger point, distal acupuncture point. Because the positive pole is known to have analgesic qualities, it is recommended that the red (+) electrode be placed as close to the point of pain as practical if the TENS unit is monophasic or has a direct current bias. Otherwise, the red (+) electrode should be proximal.

microamperage range, when bone is stressed. Conversely, application of electrical energy to bone enhances osteogenesis, again measured in extremely low amplitude ranges. TENS fits into this picture nicely, offering convenient size, shape, electrical parameters, and low cost, as compared with the existing devices for this purpose. The microampere range recommended may be obtained from the milliampere-range TENS units through the natural attenuation of current between electrodes and intervening tissues, inductive capacitance, and variable resistances. Careful selection of placements for electrodes is of prime importance and can spell the difference between success and failure.

Selecting the Parameters

The parameter selections listed below are designed to obtain the maximum "on" time for current flow and energy absorption.

1. Frequency/rate should be the highest frequency available on unit (120 Hz).
2. Pulse width should be the widest width (300 μs).
3. Intensity should be the lowest possible, "barely sensed" by patient.
4. Regimen should consist of 1 hour per session, four times daily (1 h qid).

Electrode Placement

If the fracture site is enclosed in a plaster cast, electrodes should be placed proximal and distal to the cast; two or four electrodes are used. Polarity is not important, since the distance from the target site negates any electrochemical effects found at the electrode surfaces contiguous with the skin (Fig. 6-8). If the fracture site is free of casting, three basic patterns are available for the clinician: (1) with two electrodes, one placed on either side of the fracture site, about 6 inches (15 cm) apart (Fig. 6-9); (2) with four electrodes, a crossed-pattern, about 6 inches apart, with the crossing point directly over the fracture site (Fig. 6-10); (3) with two electrodes, a "sandwich" pattern, with the fracture site between the two electrodes (e.g., dorsal and ventral) (Fig. 6-11).

Because electrode placements are not at acupuncture sites or other specific points, they may be moved occasionally, within a short distance, to avoid skin irritation if it becomes a problem with prolonged use. General electrode placement sites should be maintained for anatomic purposes, however.

Follow-Up Treatment

I suggest taking follow-up x-rays at 4 or 6 weeks to monitor changes in the status of the nonunion. Treatment should continue for at least 6 months

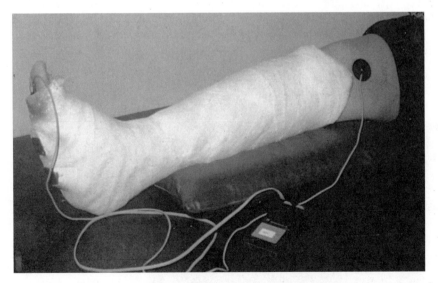

FIG. 6-8. TENS placement for a nonunion fracture of the tibia, cast present. Distal electrode(s) placed within the cast at the foot. (Miniceptor II, Med General Co., Minneapolis, Minnesota.)

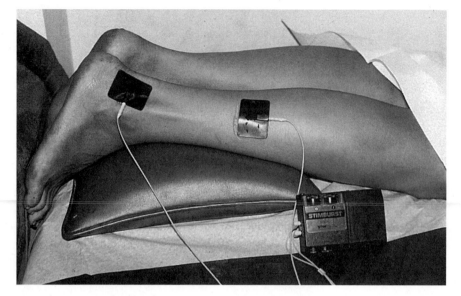

FIG. 6-9. Linear placements of TENS for a nonunion of the lower leg with no cast present. (Courtesy of Stimtech ''Stimburst,'' Randolph, Massachusetts.)

before it is discontinued. With manufactured nonunions (e.g., osteotomies and grafts), it may take more than 1 year of TENS treatment to produce favorable effects. If infection exists, progress may not occur at all until the infection is cleared.

Side Effects

The only side effects with TENS have been favorable. I have found that despite the unorthodox electrode placements, analgesia was reported by patients who had complained of pain as a symptom of the nonunion.[23] This again is a possible argument in favor of the endorphin concept.

Obstetric Cases: Labor and Delivery

The use of TENS as a form of analgesia for delivery is growing rapidly. The literature, although mostly from foreign sources, speaks highly of this procedure and has spurred American researchers to investigate this apparently safe, noninvasive, nondrug method of providing a relatively pain-free delivery.[4,5,24–27] The reluctance to use any new method with pregnant patients is understandable. Most American mothers who use TENS for delivery also participate in Lamaze programs, a favorable combination. No known or reported untoward effects are listed in

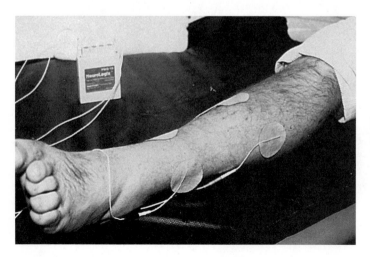

FIG. 6-10. Crossed TENS pattern for nonunited fracture of midshaft tibia.

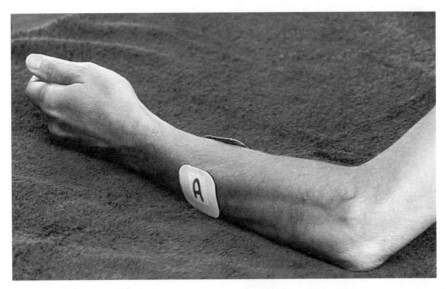

FIG. 6-11. "Sandwich" placement for nonunion fracture of the forearm, with the fracture site directly between the electrodes.

the available literature. Apgar scores are excellent.

Technique During Labor

Prior to labor, orient the mother-to-be in the use of the TENS unit and the parameter controls. Instruct her to place (or have placed) two electrodes at the level of the brassiere strap, one on each side at the spine, near the nerve roots. Elongated electrodes (1 × 6 inches [2.5 × 15 cm]) are preferred so that several nerve roots may be covered, extending distally from approximately T8 to L1 (Fig. 6-12). This circuit is activated on the morning of imminent delivery and left *on* all through delivery.

Parameters

1. Frequency should be high (80 to 120 Hz).
2. Pulse width should be medium (150 μs).
3. Amplitude should be comfortable, but low.

During Labor Contractions

A second pair of electrodes, also elongated, may be placed paravertebrally along the lowest portion of the spine without the patient sitting on them (approximately at S1 and below). Activate this circuit with each labor contraction. Parameters are the same as above, except that amplitude may be increased to block the contraction

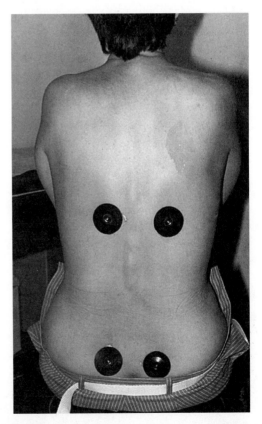

FIG. 6-12. TENS applied for labor and delivery. Proximal electrodes placed at approximately T8 (brassiere-strap level), paravertebrally, and left "on" throughout labor and delivery. Distal pair are activated with each contraction.

pain but should not be left on when the pain abates.

Second Stage of Labor: Presentation

Remove the distal two electrodes from the sacral region and relocate them to the anterior abdomen, in a V configuration, diagonal and lateral to the pubic triangle; activate them to coincide with contraction pain. Do not change frequency or width, but amplitude may be increased to block contraction pain although not enough to cause muscle contractions (Fig. 6-13).

Special Circumstances Following Labor

TENS may be used following delivery for postpartum pain. With patients undergoing cesarean section, the incision and scar discomfort may also be modified with TENS applications. The placement of electrodes is different: Lace the proximal pair as described previously, but place the distal pair at both ends of the incision or scar.

Parameters

1. Frequency should be high (80 to 120 Hz).
2. Pulse width should be medium (150 µs).
3. Amplitude should be minimal.
4. Regimen should be 1 hour, four times daily.

If pain is severe, additional sessions are recommended; if treatment is necessary for a prolonged period, *modulation* may be added to the current to avoid accommodation.

TENS for Morning Sickness and Chemotherapy

TENS is effective in the treatment of the nausea of pregnancy ("morning sickness"). It may also be administered effectively to minimize or control the nausea associated with chemotherapy (Fig. 6-14).[28] It is not effective, however, with the typical "motion sickness" nausea, since the etiology there is within the middle ear and not the general system.

Parameters

1. Frequency should be high (80 to 120 Hz).
2. Pulse width should be medium (150 µs).
3. Amplitude should be minimal but sensed.
4. Regimen should be 30 minutes every morning.

Electrode Placements

Place one electrode on the tip of the right acromion and the other electrode on the right hoku point. This technique does not work if electrodes are placed on the left extremity.

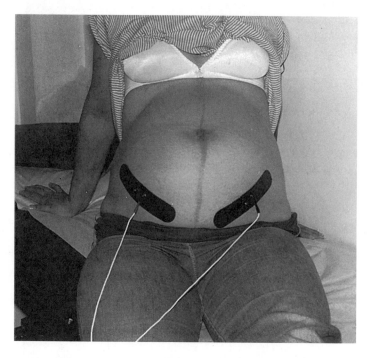

FIG. 6-13. Second-stage electrode placements.

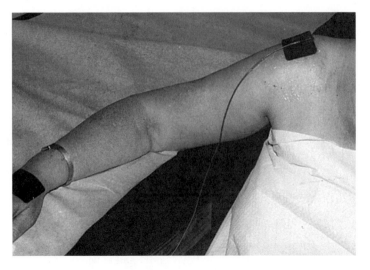

FIG. 6-14. Electrode placements for morning sickness and other forms of nausea (not otogenic). The electrodes must be placed on the right side only, at the acromial tip and hoku.

Incisional and Scar Pain

Incisional and scar pain should be treated as acute pain.

Parameters

1. Frequency should be high (80 to 120 Hz).
2. Pulse width should be medium (150 μs).
3. Amplitude should be minimal but sensed.

Electrode Placements

Place electrodes parallel to and approximately 1 inch (2.5 cm) from the incision or scar, using elongated electrodes, if available (1 × 5 in. [2.5 × 12.5 cm]) (Fig. 6-15), or in a crossed pattern, using four standard square or round electrodes (Fig. 6-16). Crossed circuits with TENS devices do not produce "net" frequencies as with interferential current techniques (see p. 79 and

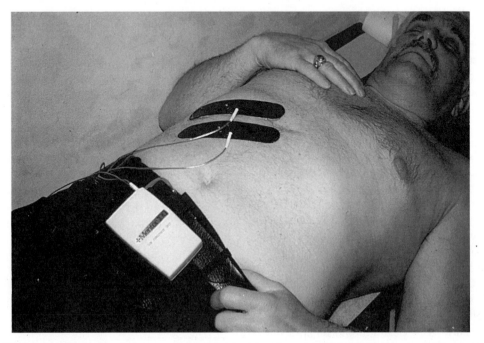

FIG. 6-15. Parallel electrode placement for painful incisional scar (gallbladder). (Mentor 100, courtesy of Mentor Corp., Minneapolis, Minnesota.)

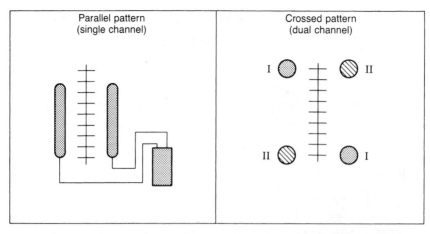

Parallel pattern
(single channel)

Crossed pattern
(dual channel)

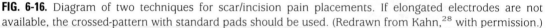

FIG. 6-16. Diagram of two techniques for scar/incision pain placements. If elongated electrodes are not available, the crossed-pattern with standard pads should be used. (Redrawn from Kahn,[28] with permission.)

Fig. 5-12). Resultant "net" frequencies are a product of the cancellation and reinforcement of waveforms from two contributing circuits differing in frequency. With crossed circuits of TENS devices, there is no cancellation, since waves are identical in both circuits. Amplification, therefore, takes place in the crossing zone, providing added stimulation in a critical region. Because of the heterogeneous nature of body tissues, and the consequent conduction of currents into neighboring tissues, there is no "crossing point," as one would expect. There is a sector within the target area that receives a varying degree of stimulation. However, it is localized within the perimeter of the electrode "square." Precise electrode placement by the clinician should determine this sector (see Fig. 3-12).

Home Use of TENS

TENS is basically designed as a unit to be used daily at home by patients. It is wise, however, to prescribe a TENS unit for use at home *only* after successful trials in the office or clinic. It is the policy in my facility to have the parameters set by the physical therapist, leaving only the amplitude (intensity) control to the patient or family. Warn patients about using the unit at sites other than those prescribed, at settings different from those set by the clinician, and on other people. After several successful sessions at my office that verify TENS as an effective pain modification device, I prescribe the identical model for the patient to use at home according to my directions.

Compliance With Instructions for TENS Use at Home

Recent innovations with TENS (and EMS) units facilitate the monitoring of compliance with clinicians' instructions. Electronic components can register and record usage at home, store the data, and display the results for the clinician at re-evaluation visits. These units can also be programmed with the identical successful parameters obtained in the office with larger, more complex clinical models.*

Advising the Patient in Home Use

1. Do not leave it to the dealers to instruct the patient in the correct use of the TENS unit; they are neither trained nor legally qualified.[29]
2. Do not order the patient a model other than the one used successfully in the office trials. This is not only unfair to the manufacturers, but often leads to results other than those predicted, since each model is somewhat different. The selection of TENS units should *not* be left to dealers, distributors, administrators, insurance carriers, or other third-party payers.
3. Keep accurate records of the patients' parameters and results.
4. Have several brands available for your prescriptions and procedures.
5. Accept responsibility for the success or failure of your work with TENS.

---

*N-4 and N-16 units, Neurotech Co., Phoenix, AZ; Ultrastim unit, Meuromedics Co., Clute, TX.

SUMMARY

TENS has been partially responsible for opening new areas of treatment not traditionally associated with physical therapists. The treatment of pain now accounts for a large part of our professional practice. Applications of physical treatment are available to all specialties, whether anatomic or pathologic. The astute clinical physical therapist will find a broad spectrum of suitable targets for TENS—unlimited in clinical scope and with opportunities for innovative techniques.[30–33,35]

REFERENCES

1. Dougherty RJ: TENS, an alternative to drugs in the treatment of chronic pain. AAFP, San Francisco, September 25–28, 1978
2. Anonymous: TENS for treatment of post-operative incision pain. US Department of Health and Human Services Assessment Report Series. Vol. 1. No. 15, 1981
3. Kahn J: TENS for non-united fractures. JAPTA 62:840, 1982
4. Augustinsson LE, Bohlin P, Bundsen P, et al: Pain relief during delivery by TENS. Pain 4:59, 1977
5. Tannenbaum J: Protocol for TENS with Labor and Delivery. Department of Electronics, Hadassah Hospital, Jerusalem, Ein Karem, Israel, April 1980
6. Kahn J: TMJ Pain Control. Whirlpool, PPS/APTA, Fall 1982, p. 14
7. Alm WA, Gold ML, Weil LS: Evaluation of TENS in podiatric surgery. JAPA 69:537, 1979
8. Melzack R, Wall PD: Pain mechanisms: a new theory. Science 150:971, 1965
9. Adler MW: Endorphins, enkephalins and neurotransmitters. Reprinted in Med Times 110:32, 1982
10. Sjoelund BH, Eriksson MBE: Endorphins and analgesia produced by peripheral conditioning stimulation, p. 587. In Bonica JJ (ed): Advances in Pain Research and Therapy. Vol. 3. Raven Press, New York, 1979
11. Rausch D: TENS. Rx: Home Care, August–September 1981, p. 81
12. Search for the body's own pain killer. In Scavina L (ed): Pain Control. Staodynamics Corp., Longmont CO, September–October 1979, No. 22
13. Shade S: Use of TENS for a patient with cardiac pacemaker. JAPTA 65:206, 1985
14. Stanton D: Polarity of Wave Forms. Medtronic Corp., Minneapolis MN, 1978
15. Mannheimer JS, Lampe GN: Clinical TENS. FA Davis, Philadelphia, 1984, p. 199
16. Gersh MR, Wolf SL: Applications of TENS in the management of patients with pain. JAPTA 65:314, 1985
17. Howson DC: Peripheral neural excitability. JAPTA 58:1467, 1978
18. Eriksson M, Sjoelund B: Low frequency TENS. Reprinted from Med Times, January 3, 1979
19. Bassett A, Pilla A, Pawluk R: A non-operative salvage of surgically resistant pseudoarthrosis and non-unions by pulsating electromagnetic fields. Clin Orthop 124:128, 1977
20. Weiss A, Parsons J, Alexander H: DC electrical stimulation of bone growth: review and current status. J Med Soc NJ 77:523, 1980
21. Anonymous: Electricity and Bone Healing. Harvard Medical School Health Letter, October 1981
22. Kahn J: TENS for non-united fractures. JAPTA 62:840, 1982
23. Kahn J: Using TENS to Facilitate Delayed Union in Fractures: Case Report. PT Advance, King of Prussia, PA, January 1996
24. Brown G, Machek O, Vivian J: TENS management of postoperative pain in obstetrical and gynecological procedures. Pain Control (abstr) No. 39. Staodynamics Corp., Longmont, CO, November–December 1983
25. Anonymous: TENS during labor and delivery. Clinical Protocol. Department of PM & R, Boulder Memorial Hospital, Boulder CO. Pain Control (abstr) No. 39. Staodynamics Corp., Longmont, CO, November–December 1983
26. Wilczynsky C, Ostman T: Pain relief in labor by TENS. Senior paper, State University of NY at Stony Brook, PT Program, presented at the NY Chapter APTA Conference, New York, April 1985
27. Keenan DL, Simonsen L, McGrann DJ: TENS for pain control during labor and delivery. JAPTA 65:1363, 1985
28. Kahn J: Clinical Electrotherapy. 4th Ed. J Kahn, Syosset, NY, 1985
29. Kahn J: Unlawful administration (Editorial). Rx: Home Care. Los Angeles, June 1985, p. 9
30. Anonymous: Electrical stimulus modulates ischemia. Medical World News, October 8, 1984
31. Kaada B, Olsen E, Eielsen O: TENS with vasodilation. Gen Pharmacol 15:107, 1984
32. Kahn J: Pain control with electrotherapy. PT Forum 4:12, 1985
33. Kahn J: The role of electrotherapy in comprehensive health care. Doctoral dissertation, Century University, 1984 (abstract only, available from author)
34. Romero-Sierra C: Neuroanatomy: A Conceptual Approach. Churchill Livingstone, New York, 1986, p. 199
35. Wang SF, Nelson C, Gross M, Light A: The effect of TENS on somato-sensory evoked potentials and subjective pain. J Clin Electrophysiol 4:23, 1992

Iontophoresis

Iontophoresis, also termed *ion transfer,* is the introduction of substances into the body for therapeutic purposes by means of a direct current. Each substance is separated into ionic components by the action of the current and deposited subcutaneously, according to the imposed polarity on the electrode. Therapeutic results depend on the ion introduced, the pathology present, and the desired effects. There is no "indication" for iontophoresis per se; the reference must be to the ion selected. Similarly, there can only be "contraindications" to individual ions, based on patient sensitivities, allergies and complicating factors in specific instances.

(see Table 7-2). Manufacturers' patent limitations must not be confused with the general purpose of this modality. Some manufacturers also recommend exclusive use of their patented, disposable electrode systems. This may add to the cost of administration and may account for the reluctance on the part of some facilities to employ iontophoresis.

In summary, practitioners should be aware of the multiple applications of iontophoresis with a wide variety of chemicals. They should also recognize the long history of this modality in the literature of our profession—utilizing apparatus quite different from today's modern array, yet producing equally successful results.

BACKGROUND

Iontophoresis has been in use for more than a half-century and was mentioned in the early literature of the late 1700s and 1800s. Abramowitsch and Neuossikine, in their classic treatment of iontophoresis,[1] listed scores of references of workers in this field prior to the advent of modern equipment, chemicals, and applications.[2]

Since the first edition of this book was published, the interest in and enthusiasm for iontophoresis have increased considerably. Several new companies have introduced instruments designed exclusively for ionic transfer as adjuncts to existing models of electrical stimulation apparatus or as individual units (Fig. 7-1). Computerization has reached the iontophoresis levels now, with units specifically designed to be programmed for this purpose (Figs. 7-2 and 7-3).

It should be remembered that the concept of iontophoresis is *not* new, nor is it meant to be utilized with any single substance. There is a long list of ions available to the practitioner applicable in a broad spectrum of pathologic conditions

Physics

The current that is required for ion transfer, continuous galvanic direct current, is obtained from standard low-voltage generators (Fig. 7-4), as well as from the battery-operated units now available (Figs. 7-5 and 7-6). Claims for ionic transfer with other types of current (e.g., high-voltage direct current) have not been substantiated and have been withdrawn by manufacturers. Substances used, for the most part, are basic elements plus several radicals of physiologic value.

Treatment is not the direct current itself but rather the ions introduced. Selection of specific substances depends on the target reaction or condition.

The completely noninvasive concept of iontophoresis is made even more attractive to clinicians because of the minimal ionic concentrations required for effective administration. The old adage of "the more the better" definitely does not hold with ion-transfer techniques. In the following section, there is no reference to the concentra-

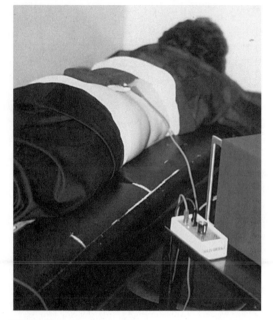

FIG. 7-1. An auxiliary continuous direct current unit connected to an Electrostim 180. (Courtesy of Electrostim, Joliet, Illinois)

tions of solutions or ointments; the advantage lies in the small percentages found in most common ion sources. Availability and costs are both affected favorably.

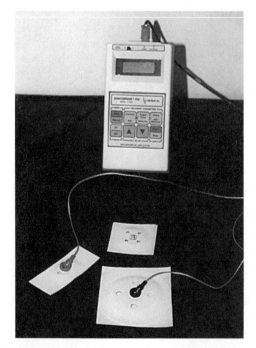

FIG. 7-2. The Iontophor—a computerized, programmable, battery-operated iontophoresis device. (Courtesy of Life-Tech Corp., Houston, Texas.)

Iontophoresis vs. Phonophoresis

Phonophoresis is discussed in detail in Chapter 4, Ultrasound. Often these two modalities are improperly equated. Iontophoresis is the introduction of ions into the body, whereas phonophoresis is molecular transmission across the skin. The question of availability for recombination to useful compounds becomes evident when the two levels of physical state are involved. Ultrasound is often more rapid in effect, whereas iontophoresis requires more time to produce results. I believe that the more rapid effects are attributable to the enhanced absorption across membranes that ultrasound causes; however, it causes that absorption at the molecular level, necessitating breakdown and recombination for lasting effects. The electrical qualities and characteristics of ion transfer are best contrasted to the soundwave-generated energy *after* study of both has been completed and clinical applications of both have been evaluated.

Formula for Iontophoresis

The basic formula for using iontophoresis is $I \times T \times ECE =$ grams of substance introduced.

Intensity *(I)* is measured in amperes and time *(T)* in hours; electrochemical equivalency *(ECE)* refers to standard figures for ionic transfer with known currents and time factors. An example is magnesium,[1] found in a solution of magnesium sulfate.

$$I \quad \times \quad T \quad \times \quad ECE =$$
$$5 \text{ mA} \quad 30 \text{ min} \quad 0.0115 =$$
$$0.005 \text{ A} \quad 0.5 \text{ h} \quad 0.0115 = 0.0002875 \text{ g}$$
$$= 0.2875 \text{ mg}$$

It is much more difficult to analyze and determine the *ECE* for many radicals and complex substances (e.g., steroids). It may be safely assumed, however, that even fewer milligrams of these larger, more complex submolecular substances will penetrate the skin.

Ionic Polarity

The basis of successful ion transfer lies in the physics principle: "Like poles repel, and unlike poles attract." Ions, being charged particles with positive or negative valences, are repelled into the skin by an identical charge on the electrode surface placed over it. Once subdermal, the ions

FIG. 7-3. A recent addition to the field is the LVG325. This unit is a direct current generator only, with continuous mode for iontophoresis and surged or alternating polarity modes for stimulation. (Courtesy of Amrex Corp., Houston, Texas.)

introduced recombine with existing ions and radicals floating in the bloodstream, forming the necessary new compounds for favorable, therapeutic interactions. Therefore, selection of the correct ionic polarity and matching it with the "like" electrode polarity for administration is of prime importance.

The use of *dexamethasone sodium phosphate with lidocaine* is recommended by several manufacturers. This powerful (and expensive) ionic source is available through the manufacturers but can also be obtained from other commercial sources.

Originally suggested as a *positive* ion, under the *anode,* recent communications from one manufacturer advise switching to the *cathode* if results are not satisfactory. The electrochemistry involved is rather complex and requires detailed explanations on the part of the manufacturer, even for "old-timers" like myself. It is still difficult for me to conceive efficacy under BOTH poles, unless, of course, the alternating polarities drive in *different* ions. Again, however, clinicians report success with this procedure. Much more is to be learned along the way.[3,4]

Low-Level Amperage

Experience and research have shown that low-level amperages are more effective as a driving force than are high-current intensities.[1,5,6] Higher intensities apparently adversely affect the inter-

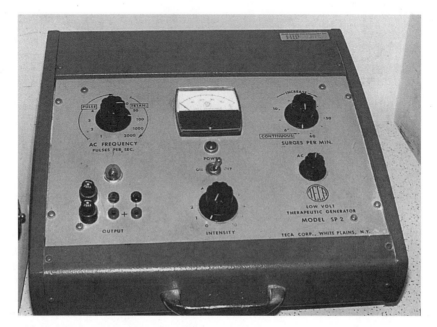

FIG. 7-4. Typical low-voltage generator unit used for iontophoresis. Selector switch is set on continuous direct current. (SP-2, courtesy of TECA Corp., Pleasantville, New York.)

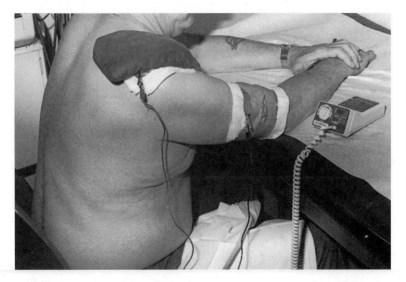

FIG. 7-5. Recently developed, battery-operated continuous direct unit for iontophoresis administered to the shoulder, with the indifferent electrode on the upper arm. (Phoresor, courtesy of Iomed Corp., Salt Lake City, Utah.)

ionic phenomena and hinder penetration.[5] This is analogous to trying to enter a crowded freeway with a multivehicle convoy at high speed; a single vehicle, at a slower speed, would find entry much less difficult. Neither the strength of the solution nor the percentage of the ointment for adequate ionic transfer *and* effect need be high, since the ions administered are active, i.e., "charged," and ready for recombination immediately. A few

ions, therefore, ready to combine, may do a better job than a multitude of ions, mutually repelling each other because of their smaller charges with fewer entering usable configurations. The physics involved with iontophoresis, then, necessitates low milliamperage (maximum 5 mA) and low percentage of ion sources (1 to 5 percent).

Enlarged Negative Electrode Importance

The negative electrode is more irritating than the positive due to the caustic sodium hydroxide formed where it is positioned. High concentration of hydrogen ions and the relative rapidity of reactions at the cathode (–) suggest a method of reducing the current density (i.e., current per square centimeter) under the negative electrode to avoid undue irritation and possible burns. This is best and most easily accomplished by making the negative electrode larger than the positive and should be done even if the negative electrode is the active, or driving, electrode. Some early texts suggested that current always flowed from the smaller to the larger electrode. This is obviously contrary to the laws of physics: Electrons always flow in the same direction—from negative to positive—regardless of electrode size. By enlarging the negative electrode, usually twice the area of the positive, the current density is lowered on the negative pad, reducing irritation effectively (Fig. 7-7). Electrochemical effects are, however, thought to be highly localized and

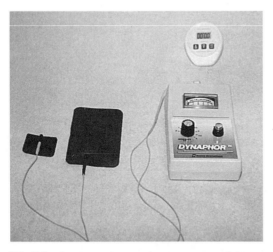

FIG. 7-6. The Dynaphor, a battery-operated iontophoresis device with direct readout for intensity, is shown here with standard carbon-rubber electrodes for pin-leads. Also shown is an included digital treatment timer. This unit is primarily designed for use with chemical-impregnated membranes that serve as ion sources (hydrocortisone, lidocaine, salicylate, acetate, etc.). (Courtesy of Henley International, Sugarland, Texas.)

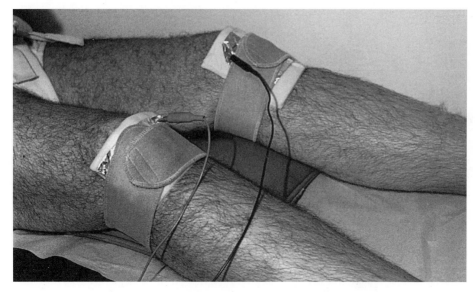

FIG. 7-7. Double iontophoresis to bilateral knees. The smaller electrode (patient's right) is positive with 1 percent hydrocortisone ointment, while the larger (patient's left) is the negative with iodine-salicylate ointment. Note that with double iontophoresis, there is *no* indifferent electrode, since *both* are active.

rarely extend farther than 1 mm from the electrode surface. The chemical effects from the introduced ions have a more profound effect and extend to deeper levels by capillary actions, reaching successively lower levels aided by the biophysical conductance of the current. The skin under the negative electrode requires additional care; this is discussed on p. 137, step 15.

PHYSIOLOGIC CHANGES

The physiologic effects of ion transfer can only be expressed in terms of the ion selected for treatment, since each is different. Specific ions are utilized for the treatment of pain, spasm, and inflammation; this is discussed in detail in the Treatment section. Some effects common to all are considered below.

Ion Penetration

Penetration is generally less than 1 mm, with subsequent deeper absorption via capillary circulation and transmembrane transport. The bulk of the ions deposited are found directly at the site of the active electrode, where they are stored as either a soluble or insoluble compound, to be depleted by the sweep of the circulating blood or utilized locally as a concentrated source for

further recombination.[1] In this manner, certain substances known to be irritants to the stomach, such as hydrocortisone or salicylates, may be introduced locally with little or no untoward reactions to the gastric mucosa.

Acid/Alkaline Reactions

The anode (+) produces an acid reaction, a weak hydrochloric acid, whereas the cathode (–) produces a strong alkaline reaction, sodium hydroxide. The anode is sclerotic and tends to harden tissue, serving as an analgesic, possibly due to the local release of oxygen, aiding in the vitality of the tissues. Conversely, the cathode is sclerolytic, a softening agent, releasing hydrogen, and is utilized clinically in the management of scars, burns, and keloids. The irritating quality of the sodium hydroxide and the rapid concentration of active hydrogen ions make the cathode ideal for the stimulating electrode with electrical stimulation procedures (Fig. 7-8).

Joule's Law

The anode and the cathode both produce hyperemia and heat because of vasodilation and Joule's law: $H = 0.24$ EIT. This relationship indicates that the conversion of electrical energy to heat is directly proportional to the intensity, resistance,

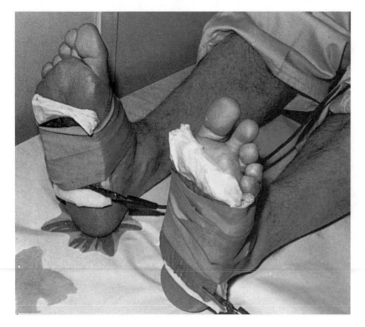

FIG. 7-8. The technique for iontophoresis for hyperhidrosis with tap water with the recommended size differential between negative and positive electrodes. Rubber bandages secure electrodes.

and time. In most instances, the reddening of the skin under both electrodes disappears within 1 hour of the treatment and should cause no alarm to patient or clinician. Cathodal (−) hyperemia is generally more pronounced and lasts a little longer than that of the anode (+).

Dissociation

Normally, ionizable substances dissociate in solution, releasing ions and radicals free to drift toward the opposite poles when a direct current is passed through the solution, following the rule that unlike poles attract.

The additional current-enhanced flow of electrons increases this normal migration and gives meaning to the concept of ion transfer or iontophoresis. Because various tissues offer different resistances to current flow, electrode placement becomes a vital factor; this is discussed in the Preparing the Electrodes section.

Pathology and Ion Selection

Ions and their relationship with pathologic conditions are listed in Table 7-1 for clinical references only. Detailed data on sources, polarity, and clinical applications will be found in the Techniques section in Chapter 9. References in the current literature list antibiotics for ion transfer,[17] as well as variations of steroids, such as

cortisol and dexamethasone. Abramowitsch and Neoussikine[1] and Kahn[7] offer a broad spectrum of iontophoretic substances for a wide range of specific effects[18] (Figs. 7-9 to 7-13).

COMPLICATIONS

Burns

Burns are not a welcome addition to any treatment protocol and are a constant consideration

TABLE 7-1. Ions and Their Relationship With Pathologic Conditions

Pathology	Ion Selection
Pain	Lidocaine, hydrocortisone, salicylate, magnesium
Inflammation[8–10]	Hydrocortisone, salicylate
Spasm[11,12]	Magnesium, calcium
Ischemia[7]	Mecholyl, magnesium, iodine
Edema[7]	Mecholyl, magnesium, salicylate, hyaluronidase
Calcific deposits[13]	Acetic acid
Fungi[7]	Copper
Gouty tophi[14]	Lithium
Open lesions[15]	Zinc
Allergic rhinitis[7]	Copper
Scars, adhesions	Chlorine, iodine, salicylate
Hyperhidrosis[16] (Fig. 7-8)	Tap water with alternating polarity
Hypo/hyperirritability	Calcium

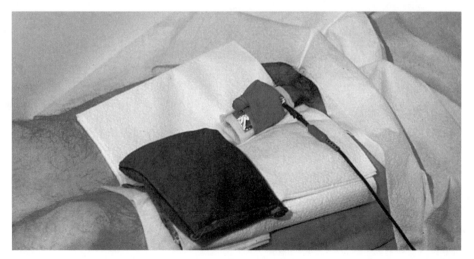

FIG. 7-9. In the management of Peyronie's disease, iontophoresis with hydrocortisone is a method of choice to reduce the inflammation in the target area. Here a small aluminum electrode is fashioned and superimposed over moist toweling. (An alternative method is discussed on p. 154 and illustrated in Fig. 9-11.)

with all electrotherapeutic modalities, iontophoresis not excluded. Most burns are directly due to poor technique on the part of the clinician.

Chemical Burns

Chemical burns usually result from excessive formation of the sodium hydroxide at the cathode. Commonly known as lye, sodium hydroxide causes a sclerolysis of tissue that takes a long time to heal. The raised, pinkish lesion generally appears immediately after treatment; hours later this turns to a grayish, oozing wound. Such wounds are best treated with antibiotics, sterile dressing, and patience. Burns under the anode are rare, since the nature of positive polarity produces a sclerotic effect, toughening the skin: a lesion that is a hardened, red area, similar to a scab. It should be treated in the same way as

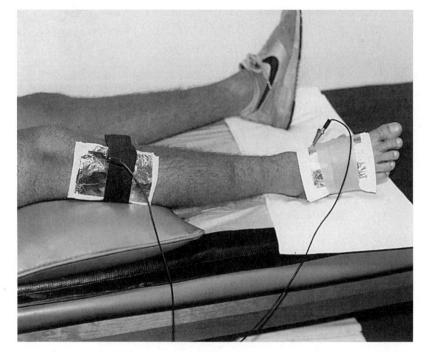

FIG. 7-10. Mecholyl iontophoresis under the positive proximal electrode provides vasodilation to the anterior tibial and peroneal distributions in a peripheral vascular deficit to the foot.

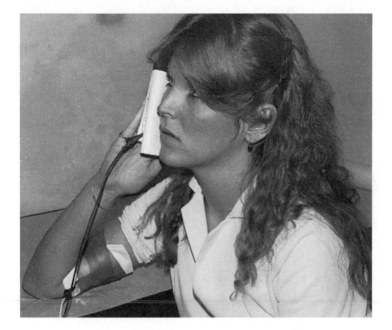

FIG. 7-11. Lidocaine iontophoresis (or hydrocortisone) in the management of pain with a temporomandibular joint syndrome.

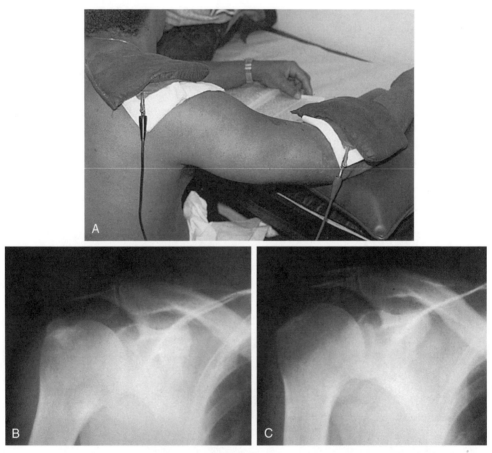

FIG. 7-12. **A,** Acetic acid iontophoresis for a calcific deposit in the deltoid region, with the indifferent positive electrode on the forearm. The acetate radical replaces the carbonate radical in the insoluble calcium carbonate deposit, forming soluble calcium acetate. **B,** Radiographic evidence of a calcific deposit in the acromial region prior to acetic acid iontophoresis. **C,** Radiograph of the same region after a series of acetic acid iontophoresis treatments. Calcific deposit has disappeared.

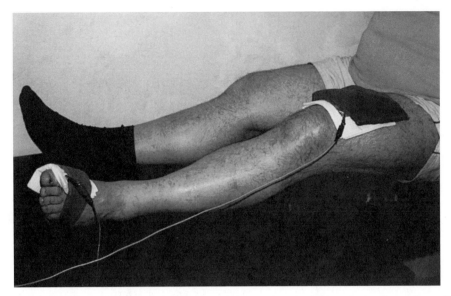

FIG. 7-13. Lithium chloride iontophoresis for gouty tophi (crystals). The lithium ion replaces the sodium ion in the crystals' insoluble sodium urate molecule, forming lithium urate, which is soluble in the bloodstream.

a chemical burn. Excessive formation of either chemical can be related directly to increased intensities of current or, more likely, excessive current densities. The traditional textbook recommendation of 1 mA per square centimeter should be disregarded. Safe levels of current will be suggested under Treatment Steps.

Heat Burns

Excessive heat buildup in areas where resistances are high will produce burns, just as will any other form of heat. High resistance is found around freckles and other sclerotic skin zones. Most high-resistance burns, however, occur when electrodes are not moist enough, when wrinkles prevent good skin contact with electrodes, or when a nonconforming, stiff electrode is placed on a rounded anatomic target site, leaving gaps of air space (high resistance) between skin and the electrode surface. Another possible source of heat burns is the result of an ischemic condition brought about by the patient's full body weight resting on an electrode, effectively "squeezing" out the available blood supply from the area and thus reducing the natural coolant quality of flowing fluid (blood). Lightweight sandbags, rubber bandages not tightly wound, and changes in position for the patient will help greatly in preventing burns from this cause. Heat burns are

treated as any type of burn, with antibiotics and sterile dressing.

Sensitivities and Allergic Reactions to Ions

Sensitivities and allergic reactions to ions are rare, yet important. In contrast to burns, which have local effects, ions have systemic effects. It is often difficult to predict which patient will be sensitive to which ions. Some practical suggestions for prevention follow:

1. If the patient is allergic to seafood, do not use iodine.
2. If the patient has had a bad reaction to an intravenous pyelogram (IVP), do not use iodine.
3. Some patients with asthma react poorly to the odor of mecholyl; black coffee and fresh air will help.
4. Although very little hydrocortisone ever reaches the stomach, patients with an active ulcer or gastritis may react poorly to its administration.
5. Patients who have problems with aspirin may also react poorly to salicylates administered with iontophoresis.
6. Patients sensitive to metals (e.g., bracelets, watchbands) may also react to copper, zinc, magnesium, and other metals.

RESULTS

In most instances, the desired results will be manifested in a reduction of pain and other symptoms of inflammation, increased ranges of motion, reduction of spasm, increased tone, or any of the conditions listed previously. Often, these results are not immediate; several, however, are fast-acting and will not cause either the patient or the referring physician to question the effectiveness of iontophoresis as a modality.[19]

TREATMENT

Selecting the Appropriate Technique

Before administering iontophoresis, the clinician must first answer the following questions:

1. What is the underlying pathology to be treated?
2. What physiologic conditions must be altered?
3. Which ion is known to be effective in this condition?
4. Are there any contraindications to this ion?
5. Is there an alternate ion?
6. What is the most convenient source for this ion?
7. What is the polarity of this ion?
8. What are the electrochemical and electrophysiologic properties of this ion?
9. What placement of electrodes will offer optimal passage of current?
10. What is the best position for the patient and target tissue?
11. What milliamperage is suggested for the patient's skin sensitivity?
12. What should be the duration of the treatment?
13. How frequently and for how long should the patient be seen?
14. What results are sought?
15. Are supportive or concurrent measures required to enhance the iontophoresis?

Only when these questions have been answered can the astute clinician properly utilize iontophoresis to the fullest advantage, avoiding the pitfalls of rote treatment. The material contained in this chapter should enable the physical therapist to answer these questions and establish an effective protocol for iontophoretic procedures.

Selecting the Appropriate Ion

The list of currently utilized ions and sources (Table 7-2) should be updated by clinicians periodically to comply with rulings and commentaries by the FDA, as should references in the literature and reports from practitioners.[7]

Some General Considerations Regarding Ion Selection

1. Most *negative* ions are sclerolytic and should be utilized when a "softening" of target tissues is desired (e.g., scars, adhesions, fibrotic nodules, adhesive capsulitis, calcifications, and myospasm).
2. The oxygen liberated at the *positive* pole is of great value in the healing process and "toughening" or sclerotic process; it is also a major factor in the analgesic characteristic of the anode.
3. Some *negative* ions are also analgesic, in addition to their sclerolytic nature; salicylate is a prime example.
4. Similarly, there are *positive* ions that also serve as a sclerolytic agent (e.g., calcium, useful with stenotic tendinitis).
5. Mecholyl and magnesium are both vasodilators. The basic difference between these two positive ions is purely clinical: mecholyl usually produces immediate vasodilation and analgesia, with a vigorous reddening of the target tissues. Magnesium does not have the immediate effect of mecholyl. However, over an extended period, the vasodilation is pronounced and lasts longer.
6. Hyaluronidase is difficult to work with and requires premixing for each administration. It has no shelf life. The clinical results with this enzyme are gratifying, making the effort and the additional expense worthwhile in the management of edema. Mecholyl serves as a reasonable substitute should hyaluronidase not be available.
7. Hydrocortisone, 1 percent ointment, has been found to be a reliable ion for anti-inflammatory action. I have not found higher percentages or other more complex and potent compounds any more effective.

Precautions

It is not advisable to use two chemicals under the same electrode, even if they are of the same

TABLE 7-2. Currently Utilized Ions: Properties and Sources

Hydrocortisone: 1 percent ointment, various local sources; positive pole; anti-inflammatory; avoid ointments with "paraben" preservatives; used for arthritis, tendonitis, myositis, bursitis

Mecholyl: Mecholyl ointment, Gordon Laboratories, Upper Darby, Pennsylvania; positive pole; vasodilator, analgesic; used for neuritis, neurovascular deficits, sprains, edema (Fig. 7-10)

Lidocaine: From Xylocaine 5 percent, Astra Pharmaceutical Co., Westboro, Massachusetts; positive pole; anesthetic analgesic; used for neuritis, bursitis, painful range of motion (Fig. 7-11)

Acetic acid: 10 percent stock solution, cut to 2 percent; negative pole; used for calcific deposits, myositis ossificans, frozen joints (Fig. 7-12)

Iodine: From Iodex (with methyl salicylate); negative pole; sclerolytic, antiseptic, analgesic; used for scar tissue, adhesions, fibrositis

Salicylate: From Myoflex (Adria Laboratories, Columbus, Ohio) ointment, 10 percent salicylate preparation, or Iodex *with* methyl salicylate (Medtech Laboratories, Inc., Cody, Wyoming); negative pole; decongestant, analgesic; used for myalgias, rheumatoid arthritis

Magnesium: From 2 percent solution of magnesium sulfate (Epsom salts); positive pole; antispasmodic, analgesic, vasodilator used for osteoarthritis, myositis, neuritis

Copper: 2 percent solution, copper sulfate; positive pole; caustic, antiseptic, antifungal; used for allergic rhinitis, dermatophytosis (athlete's foot)

Zinc: From zinc oxide ointment 20 percent; positive pole; caustic, antiseptic; enhances healing; used for otitis, ulcerations, dermatitis, other open lesions

Calcium: From calcium chloride, 2 percent solution; positive pole: stabilizer or irritability threshold; used for myospasm, frozen joints, trigger-fingers, mild tremors (non-Parkinsonian)

Chlorine: From table salt (NaCl), 2 percent solution; negative pole; sclerolytic; used for scar tissue, adhesions

Lithium: From lithium chloride or lithium carbonate, 2 percent solution; positive pole; specifically for gouty tophi (Fig. 7-13)

Hyaluronidase: From Wydase (Wyeth, Philadelphia, Pennsylvania); solution to be mixed as directed on vials; positive pole; absorption agent for edema, sprains

polarity. Mutual repulsion may diminish the penetration sought. Possible exceptions to this are the use of Iodex with methyl salicylate, in which the iodine and the salicylate radical are bound in a common matrix.

It is *not* advisable to administer ions with opposite polarities during the same treatment session. The second ion administered tends to reverse the initial deposits and may lead to unwanted synthesis of possibly antagonistic ions. When the characteristics of both ions are desired, we suggest using each on alternate treatment days.

A suggested possibility of toxicity from the aluminum foil electrodes has been clearly refuted. *Harvard Medical School Newsletter* notes that aluminum is an extremely difficult substance to introduce into human skin.[20] With my recommended electrode system, additional insulation is provided by the many layers of moist paper towels, further minimizing this possibility.

Preparing the Electrodes

I have not found commercial electrodes to be advantageous with iontophoresis. They are generally too thin, too small, and too close together (if linked by the manufacturer), and, often, there is too little of a size differential between the anode and the cathode. Also, since many of the ion vehicles are *ointments,* some of which are *negative* ions, it becomes difficult to utilize the manufacturers' electrodes efficiently. Several brands require hypodermic syringes to inject the chemical solution into an electrode reservoir. If an ointment is used, this would necessitate using tap water in the reservoir. Should the ion be *negative,* there may be a problem in changing the polarity to have the negative electrode the larger of the two. In recent months, one manufacturer has offered electrodes of equal size based on buffering chemicals to minimize the buildup of either the hydrochloric acid at the anode interface and the sodium hydroxide at the cathode interface. It remains to be seen whether the buffering chemistry interferes with desired ionic transfer or minimizes burns.[21] In addition, the disposable commercial electrodes usually are quite expensive. Suggested methods of accommodating to the expenses include adding to the fee (charging the patient) for these electrodes. With most current fees fixed (e.g., Medicare, Medicaid, Workers Compensation, No-Fault) it is not possible to obtain additional remuneration, making the administration of iontophoresis too expensive or undesirable. Although many clinicians do use the commercial pads successfully, I much prefer to fabricate my own with inexpensive, disposable materials (Figs. 7-14 to 7-27).

Electrodes may be fabricated from paper towels, washcloths, or other absorbent materials over which household aluminum foil (heavy duty)

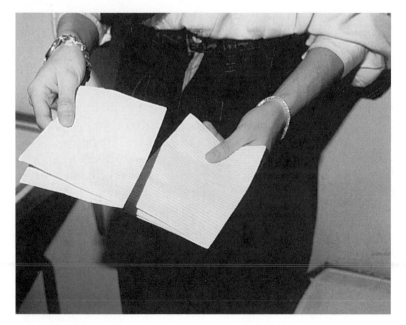

FIG. 7-14. Two paper towels for each electrode when used with iontophoresis. One towel for electrical stimulation.

is superimposed to form an electrode unit. The aluminum foil should be of several thicknesses, folded to size, rolled flat, and trimmed to be slightly smaller than the towel to avoid metal/skin contacts. The towels (two or three) should be folded wrinkle-free and thoroughly soaked in warm water or the ionic solution. Electrode units are secured in position on the patient with soft rubber bandages or lightweight sandbags (Fig. 7-28).

Positioning the Patient

Patients should never lie with their full body weight on an electrode. This creates pressure and an ischemic condition that can lead to a burn, since the cooling effect of the circulatory sweep is missing. Shoulder, hand, face, neck, elbow, and brachial regions are better treated with the patient in the sitting position (Fig. 7-28). Prone, supine, and side-lying positions are appropriate

FIG. 7-15. For the anode with iontophoresis, the two towels are folded in half so that the cathode (unfolded) is twice the size of the anode.

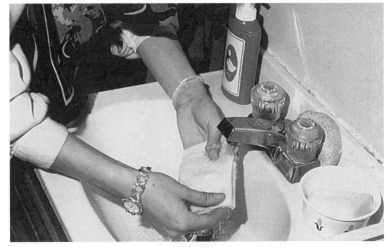

FIG. 7-16. The towels are thoroughly soaked in warm water, ensuring that there are no dry spots on the inner folds.

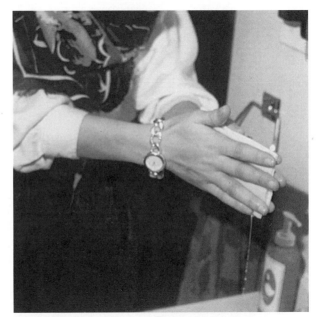

FIG. 7-17. Excess water is pressed out to eliminate unwanted wrinkles on the towel.

FIG. 7-18. Ordinary household heavy-duty aluminum foil is used to make the actual electrode.

FIG. 7-19. The aluminum foil is folded until the size desired is reached. The anode, again, should be half the size of the cathode.

for trunk and lower back, thigh, anterior chest, and abdominal procedures (Fig. 7-29). The knee is best treated with the patient in the sitting position (on the plinth, popliteal supported), as are the leg, ankle, and foot (Fig. 7-30).

Once the patient is in proper position, the following list for general procedures is recommended[22]:

Treatment Steps

1. Set all dials to zero.
2. Set selector switch to continuous galvanic current direct current.
3. Set polarity switch to correct polarity of active electrode(s) (i.e., same polarity as ion to be introduced).
4. Massage ionic ointment into the target zone, and place two folded, warm-water–soaked towels over this zone.
5. Place aluminum foil electrode pad on the towels, avoiding skin-to-metal contacts.
6. If an ionic solution is used, place two towels, soaked in the solution, over the target zone, together with the aluminum foil pads, as above.
7. With either of the above procedures, place two warm-water–soaked towels at a distant location to serve as the indifferent electrode. Remember to maintain the negative electrode at approximately twice the size of the positive, as discussed previously, regardless of which is the active electrode! Aluminum foil electrodes are superimposed on these double towel pads, as in step 5.
8. Secure electrodes in position with soft rubber bandages—not gauze, which when wet will conduct current. Lightweight sandbags (e.g., 1 or 2 pounds [0.5 or 1 kg]) are also used for larger, flatter areas of body surface. Note that sandbags moist from usage will also conduct current and should *not* be placed across two electrodes. The same warning applies to "Hydrocollator" packs used to hold electrodes in position.
9. Connect lead wires to the aluminum foil pads, using "alligator" clips (Fig. 7-31) and

Text continued on page 137

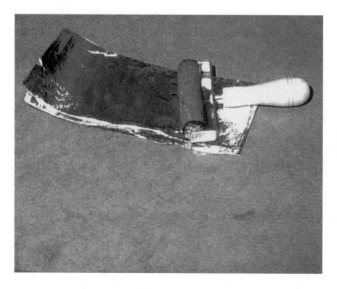

FIG. 7-20. The electrode should be rolled flat to eliminate wrinkles.

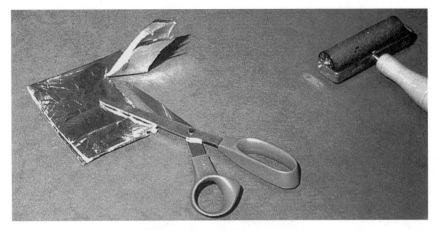

FIG. 7-21. Cut with a scissors to the desired size. Care should be taken to trim the aluminum slightly smaller than the towels to avoid overlap and skin contact.

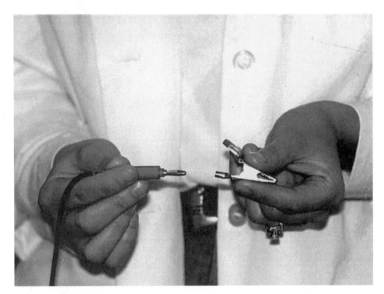

FIG. 7-22. Alligator clips should be attached to the lead wires to grip the aluminum.

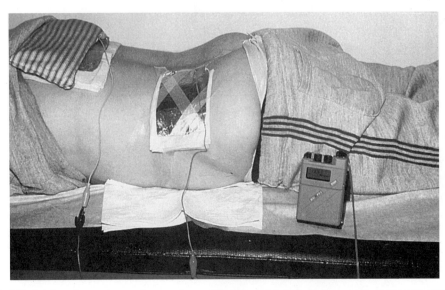

FIG. 7-23. When placed on the patient, color coding is suggested (red = positive). Secure with ordinary mending tape or lightweight sandbags. (Courtesy of EMPI Corp., St. Paul, Minnesota.)

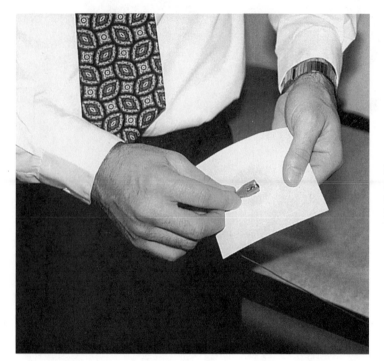

FIG. 7-24. Specialized commercial electrodes with "snap-type" electrode connectors. (Courtesy of EMPI Corp., St. Paul, Minnesota.)

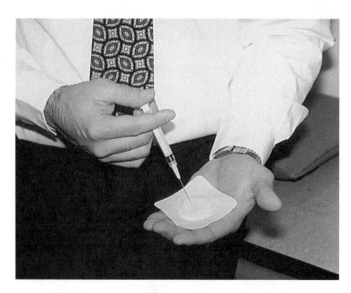

FIG. 7-25. Injection of chemical solution into electrode reservoir pad. (Needle is not necessary; the syringe will do.) (Courtesy of EMPI Corp., St. Paul, Minnesota.)

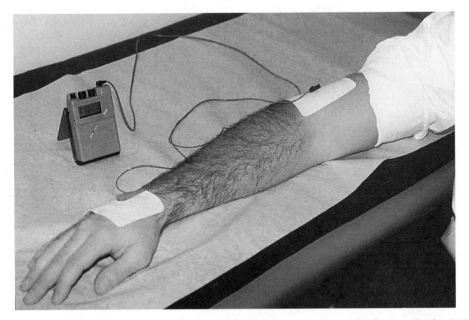

FIG. 7-26. Special electrodes are in place. The size differential here is not easily discerned. The EMPI Corp. (St. Paul, Minnesota) utilizes "buffered" pads that are designed to minimize the acid and alkali buildup at the interfaces.

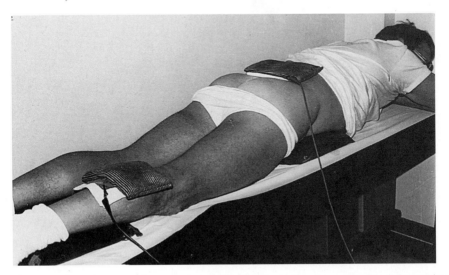

FIG. 7-27. In the iontophoretic approach to sciatic neuritis, the electrodes are secured by lightweight sandbags.

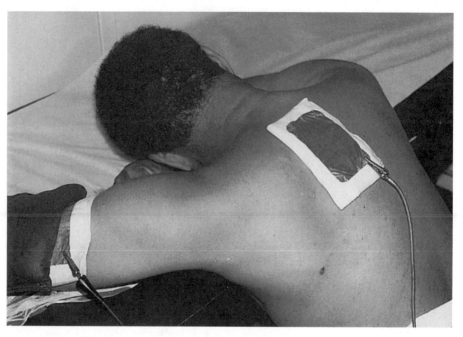

FIG. 7-28. A scapular condition is treated by iontophoresis (sandbag removed for illustrative purpose), with the patient in the sitting position, leaning forward onto a pillow.

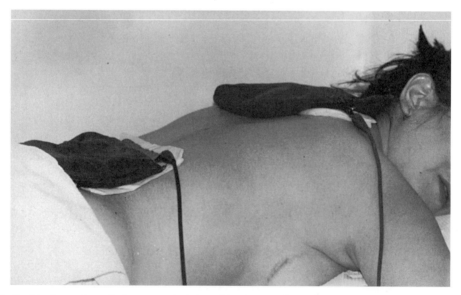

FIG. 7-29. Double iontophoresis for a cervical-dorsal-lumbosacral condition, with ions of opposite charges under each electrode. The patient is positioned prone for comfort and convenience.

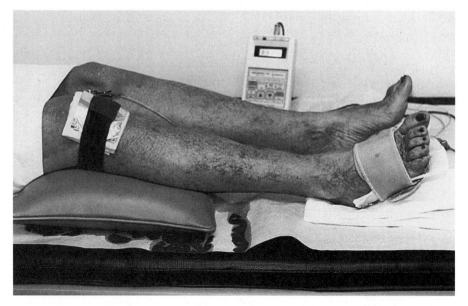

FIG. 7-30. The knee is best treated in the sitting position with popliteal support. The plantar surface is recommended for the reference electrode because of the unusually tough skin there, minimizing irritation.

bending the connecting corner away from the skin to avoid contact.

10. Turn on generator and slowly advance the control until 5 mA is reached on the direct current milliammeter.

11. The patient should feel a slight tingle, but not a burning sensation. If the patient is uncomfortable, reduce current to 2 or 3 mA.

12. Monitor the milliammeter frequently; adjust as needed to maintain at desired intensity.

13. Fifteen to 20 minutes of treatment is usually sufficient at 5 mA. If it is necessary to reduce the intensity to 2 or 3 mA, an additional 5 or 10 minutes is indicated, unless skin irritation warrants shorter treatment time. At any time, if skin irritation or a burning sensation is reported by the patient or is noted by inspection, turn off the generator slowly and check the skin. (Always turn the generator on and off slowly to avoid undesired muscle contraction by the "make/break" phenomenon of the interrupted current flow.)

14. At the termination of the treatment, turn down the current slowly, disconnect electrodes, and remove all towels from the patient.

15. Apply appropriate care to the area under the negative electrode and, if necessary because of slight irritation, to the area under the positive electrode. Because the negative is the more irritating of the two, following treatment it is advisable to apply a soothing lotion or astringent preparation to this area where no chemical (other than tap water) has been. A witch-hazel preparation,* carbolatedpetrolatum, and a light dusting with talcum powder are recommended for this purpose.

16. Follow with other modalities or procedures, with the exception of the whirlpool bath. The increased circulation from the whirlpool may tend to disperse the introduced ions more than desired. Shortwave diathermy is an

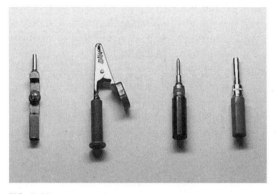

FIG. 7-31. Typical electrode tips *(left to right):* Macintosh, alligator, telephone, and banana.

---

*Köl, Nortech Laboratories, Hicksville, NY.

excellent choice of an ancillary modality, since the deeper heating tends to "draw" the ions deeper into the underlying tissues.

17. Double iontophoresis (i.e., chemicals with opposite polarities under both electrodes) enables the clinician to treat two separate locations simultaneously. With this procedure, however, there is no indifferent electrode, since both are active. The previously suggested management of the negative pole electrode is still recommended (Figs. 7-12, 7-28, and 7-30).

COMMENTS, OBJECTIONS, AND RECOMMENDATIONS

You will note that I have not included dexamethasone (sodium dexamethasone phosphate, with or without lidocaine) among the ions recommended. I have not found it to be any more effective than 0.5 to 1 percent hydrocortisone, which is less expensive and readily available over the counter. One of my objections to dexamethasone is the uncertainty regarding correct polarity. When it was first introduced by manufacturers as the *only* ion recommended with their apparatus, it was to be placed under the anode, with fair results as an anti-inflammatory agent. Shortly thereafter, practitioners were advised to place it under the cathode instead, again with fair results. This discrepancy about polarity made the actual action of this chemical questionable. Manufacturers answered with facts regarding polarity changes and alterations in the normal molecular processes once the agent was within the tissues.

My main objection, however, is to the concept that there is a single ion to treat everything. Certainly, the clinical reputation of acetic acid iontophoresis with calcific deposits, mecholyl for ischemia, magnesium for muscle spasm, salicylate and iodine for sclerolytic purposes, and, of course, hydrocortisone as an anti-inflammatory agent indicates the need for other special ions for specific conditions. Ibuprofen serves as an excellent nonsteroidal anti-inflammatory when indicated, in addition to copper, zinc, and calcium ions for a wide variety of conditions. Dexamethasone cannot do it all. Patients may be obtaining limited anti-inflammatory results with the dexamethasone *without* the specific chemical (ionic) reactions needed for tissue repair, circulatory enhancement, sclerolysis of scar and fibrotic tissues, and improved neuromuscular function.

Clinicians must assay the advantages of using ions with known polarities other than dexamethasone in these instances. Treatment plans may be designed to include *both* the dexamethasone and alternate ions on successive days (not simultaneously, however, for obvious polarity reasons).

The traditional recommendation to enlarge the cathode in relation to the anode is also being considered by manufacturers of electrodes. The highly irritating NaOH formed at the interface of the cathode with skin is minimized by increasing the surface area of the electrode to reduce the current density, thus lessening the irritation and chance of burns. Buffering of the electrode surface is claimed to eliminate the necessity for size differentiation.

When I asked whether the buffering mechanism (chemical) would interfere with the treating ions, I was told by the manufacturers that this was not found in their tests. However, I believe the testing was done *only* with dexamethasone and not the large number of other ions. I continue to recommend the size differential.

The smooth, ripple-free current available from the compact, battery-powered direct current units is ideal for iontophoresis. I find, however, that the electrodes are much too small for large anatomic areas, too thin, and often tethered together as a unit with not enough distance between anode and cathode. Current tends to flow just subcutaneously between electrodes that are too close to each other, often leading to irritation or burns along the opposing electrode edges.

Billing for iontophoresis under current insurance systems is another problem for which I have no answer other than that I administer iontophoresis when indicated, knowing I will not be reimbursed for it. However, since I am a firm believer in a multiple modality approach, I generally administer sufficient other procedures or modalities that *are* reimbursable and obtain a reasonable fee for my efforts. I continue to add iontophoresis to my treatment plan and note this on the insurance forms, since I want the gatekeepers of managed care to know that it *is* being used. My comments, objections, and recommendations are based on almost 50 years of successful clinical experience and not on extensive research. There are ample references in the literature that bear out the effectiveness of various ions in the treatment of a wide range of pathologic conditions. My training at Ithaca College (1949–1952) featured a course in electrotherapy second to none, taught by the author of the classic text on iontophoresis, Dr. David Arnon (Abramowitsch). Unfortunately, his book, *Treatment by Ion Transfer* (Grune & Stratton, 1946), is out of print but should be sought as a treasure and repository of iontophoretic information, data, and references, especially considering the year.[23–29]

REFERENCES

1. Abramowitsch D, Neoussikine B: Treatment by Ion Transfer. Grune & Stratton, NY, 1946 (out of print)
2. Haggard HW, Strauss MJ, Greenberg LA: Copper, electrically injected, cures fungus diseases. Reprinted in Science Newsletter, May 6, 1939
3. Sabbahi S, Nelson A: Iontophoresis of lidocaine/dexamethasone improves function and suppresses pain in TMJ patients, abstracted. Phys Ther 72(suppl):1992
4. Costello C, Emran A, Sabbahi M: Penetration of lidocaine into the deep tissues with iontophoresis, abstracted. Phys Ther 72(suppl.):1992
5. Mandleco C: Research—Iontophoresis. Institute for Bio-Medical Engineering, University of Utah, Salt Lake City, 1978
6. Jacobson S, Stephen R, Sears W: Development of a new drug delivery system (iontophoresis). University of Utah, Salt Lake City, UT, 1980
7. Kahn J: Clinical Electrotherapy. 4th ed. J Kahn, Syosset, NY, 1985
8. Bertolucci LE: Introduction of anti-inflammatory drugs by iontophoreses: a double-blind study. JOSPT 4:103, 1982
9. Kahn J: Iontophoresis with hydrocortisone for Peyronie's disease. JAPTA 62:995, 1982
10. Harris HR: Iontophoresis—clinical research in musculoskeletal inflammatory conditions. JOSPT 4:109, 1982
11. Kahn J: Calcium iontophoresis in suspected myopathy. JAPTA 55:276, 1975
12. Kahn J: Iontophoresis and ultrasound for post-surgical TMJ trismus and paresthesia. JAPTA 60:307, 1982
13. Kahn J: Acetic acid iontophoresis for calcium deposits. JAPTA 57:658, 1977
14. Kahn J: Lithium iontophoresis for gouty tophi. JOSPT 4:113, 1982
15. Cornwall MW: Zinc oxide iontophoresis for ischemic skin ulcers. JAPTA 61:359, 1981
16. Kahn J: Tap-water iontophoresis for hyperhidrosis. Reprinted in Medical Group News, August 1973
17. LaForest NT, Confrancisco C: Antibiotic iontophoresis in the treatment for ear chondritis. JAPTA 58:32, 1978
18. Kahn J: Iontophoresis in clinical practice. Stimulus (APTA-SCE) 8, May 1983
19. Kahn J: Iontophoresis: practice tips. Clin Management 2:37, 1982
20. Harvard Medical School Newsletter. 13:4, April 1988
21. Kahn J: Chemicals used in iontophoresis. Phys Ther Adv August 24, 1992, p. 5
22. Kahn J: Iontophoresis (video tape). AREN, Pittsburgh, PA, 1988
23. Hamill B: Iontophoresis Comes of Age. Pt/OT Today, Valley Forge, PA, October 9, 1995, p. 8
24. Balogun JA, Abidoye AB, Akala EO: Zinc iontophoresis in the management of bacterial colonized wounds; a case report. Physiother Can 42:147, 1990
25. Costello CT, Jeske AH: Iontophoresis: applications in transdermal medication delivery. Phys Ther 75:554, 1955
26. APTA: Electrical Stimulation Anthology. Vol 4. APTA, Alexandria, VA, 1993
27. Kahn J: Iontophoresis . . . red hot or just an ember? Phys Ther Adv December 16, 1996, p. 3
28. Kahn J: Iontophoresis dissected. Biomechanics, February 1996, p. 81
29. Kahn J: The chem lab. Empire State Phys Ther Albany, NY, January 1996

SUGGESTED READINGS

Gangarosa JP: Iontophoresis for surface anesthesia. JADA 88:125, January 1974

Garzione JA: Salicylate iontophoresis as an alternative treatment for persistent pain following surgery. JAPTA 58:5, 1978

Tannenbaum M: Iodine iontophoresis in the reduction of scar tissue. JAPTA 60:792, 1980

8

Testing Procedures

Although this book is primarily concerned with the therapeutic applications of electrical modalities, some mention is due the traditional methods of using electrical phenomena as diagnostic tools. These methods include alternating current and direct current manual testing, strength-duration curve determination, and chronaxie. The advent of the electromyelogram (EMG) as a diagnostic procedure in recent years has all but eliminated these three methods of electrical testing. In defense of the historical and practical values offered by the latter procedures, the rationales and techniques involved are briefly summarized. Another valuable procedure only briefly noted in this chapter is the *nerve conduction velocity (NCV) test*. Here, the rate of transmission along neural pathways is measured electronically. Results are then compared with known norms, allowing diagnostic and prognostic determinations. While the EMG is used to assess the status of neuromuscular components, the NCV test offers the practitioner additional specific neurologic data upon which to properly diagnose existing conditions. These two tests are often administered as a dual procedure. Because of space requirements and this book's intent, the EMG is not discussed here; I refer you instead to an excellent text on this subject.[1]

ALTERNATING CURRENT AND DIRECT CURRENT

General Facts

Electrical stimulation generally refers to surface electrode techniques, which, although targeted at muscles, in reality stimulate superficial nerve fibers. In turn, these fibers lead to nerve trunks and the motor points in the muscles.

1. Because normal musculature will respond to electrical stimulation by either alternating or direct current, AC or DC is simplified.
2. A poor or absent response to alternating current stimulation indicates the presence of neural damage, i.e., reaction of degeneration (RD).
3. If this occurs, direct current is preferable.
4. Vermicular ("wormy") or sluggish responses with direct current indicate moderate to severe damage to the nerve.
5. Absent responses with direct current usually signal nerve section, termed *absolute RD*.

Criteria for Presence of Reaction of Degeneration

1. Response to alternating current is absent.
2. Response to direct current is sluggish.
3. The motor point is apparently displaced distally (apparently since the motor point is a fixed anatomic landmark—the junction of the nerve with the muscle at the endplate), indicating RD. With damage, the motor point becomes desensitized and the greatest response is usually found at a point somewhat distal to the true motor point, where the bulk of fibers are found.
4. Ability to accommodate slow-rising currents is lost; since with damage the chronaxie (length of time a current must be employed to elicit a contraction) is increased, a slowly rising current will not be accommodated, as it would be in a normal neuromuscular system.
5. There is greater response at the anode than at the normally hypersensitive cathode, probably because of the polarization and repolarization phenomena at the membrane surface of the nerve fibers.
6. Higher intensities of current are required (common in the presence of RD).

Testing Technique

The following procedure is available to the clinician if a low-voltage generator with both alternating and direct current circuits is available, along with an interrupt-key stimulating electrode (Fig. 8-1).

1. Test normal side first.
2. With alternating current, place large indifferent pad, moistened with warm tap water, at a distance from the testing site and not on an antagonistic muscle group.
3. With the interrupter-key electrode, stimulate the target muscle's normally located motor point approximately once each second, using just enough current to elicit a contraction without pain (Fig. 8-2).
4. Note the presence or absence of response and the quality of the response, if present. It should be brisk.
5. A brisk response with alternating current lessens the possibility that RD exists, however.
6. Do not assume normal status based on alternating current response within 10 to 14 days of injury, because RD can take as long as 14 days to develop and manifest itself clinically. Retesting at weekly intervals is suggested.

7. If alternating current does not evoke response, change to direct current and repeat the above procedure with the stimulating electrode as the negative pole. Less current will be needed since the negative pole electrode is usually more irritating because of the alkaline chemical effects produced.
8. A brisk response may indicate only partial or slight damage to the nerve.
9. Sluggish or vermicular contractions indicate moderate to full RD.
10. Lack of response to direct current many indicate nerve severance.
11. Check results with the standard criteria above for confirmation of clinical findings.
12. Several critical responses, rather than a single or isolated finding, should be considered for diagnostic reports.

Signs of Regeneration

1. The polar formula is reversed (i.e., negative response is greater than positive).
2. Reduced intensities are required.
3. Brisker contractions are elicited with minimal current.
4. Return of motor point to traditional site is apparent.
5. Possible response to alternating current is noted.

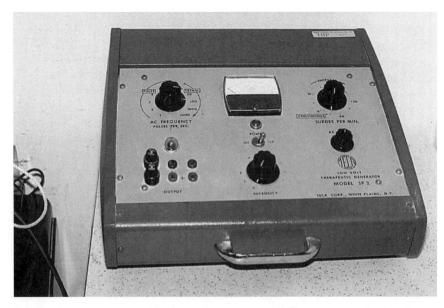

FIG. 8-1. SP-2 low voltage generator, with alternating current and direct current circuits. (Courtesy of TECA Corp., Pleasantville, New York.)

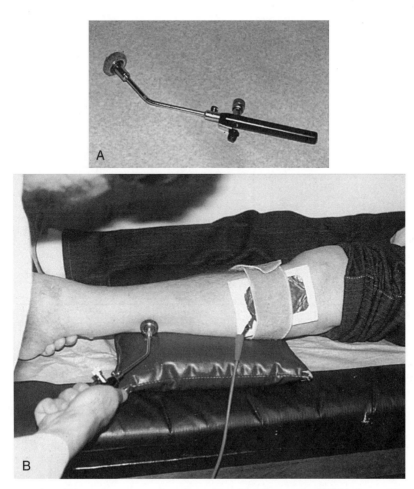

FIG. 8-2. Interrupter key **(A)** used to test peroneal muscles **(B)**. (Courtesy of TECA Corp., Pleasantville, New York.)

Value

With the above technique, it is possible to report the presence of RD to the referring physician in less than 15 minutes, with reasonable accuracy and little expense. Further evaluative tests such as chronaxie, strength-duration curves, conduction velocities, and EMGs will confirm, identify, and locate the damage more accurately. Proper diagnosis and treatment can follow; retesting at intervals may aid in establishing a prognosis.

STRENGTH-DURATION CURVES

General Facts

The following delicate test plots, in graphic form, the strength of stimulating currents against the duration of the stimuli. The accepted, traditional rule is "the stronger the current, the shorter the duration necessary to produce a contraction."

Selected durations, measured in milliseconds and fractions of milliseconds, are utilized to stimulate muscles through the nerve at motor points. The milliamperage required at each duration is then plotted on a graph, developing a mathematical curve indicating normal or abnormal conditions as compared with known norms (Fig. 8-3). Subsequent testing usually shows the curve "shifting" to the left or right, depending on whether improvement or worsening is occurring. A shift to the *left* indicates improvement, since shorter durations with more current is noted; a shift to the right indicates an increase in duration, with less current.

Equipment

Specialized apparatus is required for the following. A chronaxie unit used (Fig. 8-4).

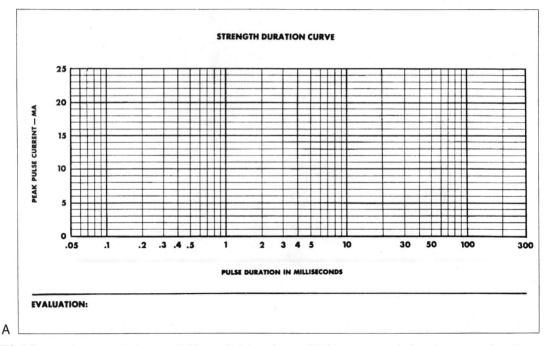

FIG. 8-3. A and **B,** Typical chart available to clinicians for establishing a strength-duration curve plot. (Form ET-3B, courtesy of TECA Corp., Pleasantville, New York.) *(Figure continues.)*

Testing Technique

1. Set all dials to zero.
2. Place indifferent electrode distant from the target muscle and not on an antagonistic muscle group.
3. Set strength-duration testing apparatus to chronaxie mode.
4. Select several durations, greater and smaller than known norms, on the millisecond dial.
5. Stimulate target motor points at each of these durations, noting on the graph the milliamperes required to elicit a minimal contraction.
6. Change the current or duration scale or both if necessary to accommodate findings.
7. Plot at least four points to establish a minimal curve.
8. Connect points with pencil to form visible curve on graph.

Results

1. The plotted curve must be compared with subsequent test graphs to determine progress or regression (see General Facts above). With chronaxie test results available, however, a simple check on results is possible. The duration plotted at twice the rheobase should fall on the curve or close to it (see Chronaxie below).
2. Retesting at weekly intervals to determine presence of shift is suggested.
3. Curves falling within normal limits indicate no damage, whereas deviations from normal limits indicate relative severity of damage to the neuromuscular system tested.
4. The shifting of the curve offers the clinician excellent prognostic evidence (Fig. 8-5).

CHRONAXIE

Often misinterpreted, the chronaxie test measures the duration of the stimulation necessary to produce a minimal contraction at specific intensities of current. It is primarily a measure of time, not intensity. A special unit required for this test (see Fig. 8-4) is available to the clinician from several manufacturers.

Normal muscles respond to short-duration stimuli, usually less than 1 ms. In RD or nerve damage, extended durations are necessary to elicit a contraction. The greater the deviation from known norms, the more extensive the damage, and the poorer the prognosis for rapid recovery.

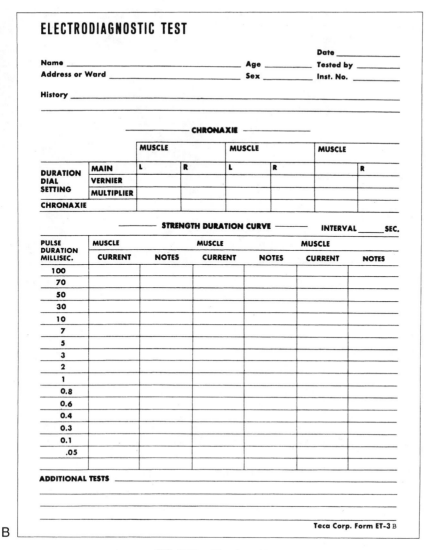

B

FIG. 8-3B. *(Continued.)*

Most texts and manufacturers provide charts with normal values for comparison studies (Table 8-1). Subsequent testing at periodic intervals (e.g., weekly or monthly) will give indications of progress or regression in a sequential form, valuable in treatment management.

Testing Procedures

Rheobase

Establishment of a base intensity to produce a minimal contraction is necessary. Most chronaxie units will offer a 300-ms duration for this procedure, since this duration is sufficient to stimulate even severely damaged nerve-muscle targets. The intensity producing the minimal contraction at 300 ms is termed the *rheobase*. In an effort to reduce the human error inherent in subjective testing, the rheobase traditionally is doubled, a procedure that effectively diminishes the percentage of error that occurs with personal observations (Fig. 8-6).

1. Establish the rheobase by setting the selector dial to rheobase.
2. Place an indifferent electrode at a distance from the test site but not on an antagonistic muscle to minimize resistance to contraction of the tested agonist.

FIG. 8-4. Unit used for both strength-duration curve and chronaxie testing. Variable pulse generator. (Courtesy of TECA Corp., Pleasantville, New York.)

3. The interrupter-key electrode may now be used to stimulate the muscle to be tested at 0.5- or 1-second intervals. With modern chronaxie equipment, however, it is possible to select the interval and place an electrode on the target site, allowing the machine to stimulate at the selected frequency mechanically.

4. When the barest of contractions is noted, throw the selector switch to chronaxie. This will automatically double the rheobase and reduce the duration from 300 ms to zero.

5. Turn the main duration control slowly in the direction of increased milliseconds, starting from zero, until a minimal contraction, similar to that elicited at rheobase, is observed.

6. Turn the main dial back one level, and turn the vernier dial up from zero to obtain a third decimal value when the minimal contraction is again reached.

7. Read off the main dial and the vernier dial to obtain the three-place decimal value for the duration established.

8. Compare this figure with the norm charts (e.g., Walthard scale) to determine comparative status of tested musculature (see Table 8-1).

Results

Deviations from the norms (i.e., increased duration values) are indicative of nerve damage. (Lowered values are not common, have no clinical significance, and generally are attributed to subjective factors.) Subsequent testing at weekly or monthly intervals is recommended for treatment management and prognosis determination.

A convenient method of checking accuracy with chronaxie testing results is to follow a strength-duration curve, performed before or subsequent to the chronaxie test, to locate the chronaxie duration at the doubled rheobase intensity on or close to the line of the curve (Fig. 8-7).

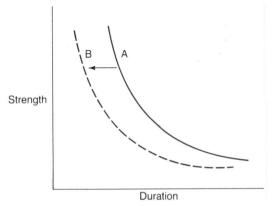

FIG. 8-5. Shifting of strength-duration curves with prognostic values. *A,* Original strength-duration plot. *B,* Subsequent plot showing shift to the left, indicating improvement.

TABLE 8-1. Normal Chronaxie Values of Muscles (Values in Sigma)

A: Upper Extremity	Walthard	Bourgignon	B: Lower Extremity	Walthard	Bourgignon
Group 1			Group 1		
Deltoid	0.04–0.29	0.08–0.16	Gluteus maximus	0.05–0.15	0.1–0.16
Biceps	0.04–0.15	0.08–0.16	Rectus femoris	0.05–0.175	0.1–0.16
Brachioradialis	0.04–0.25	0.08–0.16	Vastus internus	0.04–0.22	0.1–0.16
Triceps, internal head	0.04–0.07	0.08–0.16	Vastus externus	0.06–0.18	0.1–0.16
Group 2			Sartorius	0.04–0.26	0.1–0.16
Triceps brach. long	0.04–0.43	0.16–0.32	Adductor longus	0.07–0.16	0.1–0.16
Triceps long, external head	0.04–0.35	0.16–0.32	Gracilis	0.15	0.1–0.16
Group 3			Tensor fasc. lat.	0.05–0.16	0.1–0.16
Extensor carpi rad. longus	0.05–0.17	0.2–0.036	Group 2		
Palmaris longus		0.06–0.14	Tibialis anticus	0.08–0.38	0.1–0.16
Flexor carpi radialis	0.04–0.24	0.2–0.36	Extensor digitorum longus	0.13–0.38	0.23–0.36
Flexor carpi ulnaris	0.05–0.26	0.2–0.36	Extensor hallucis longus	0.36	0.24–0.36
Flexor sublimis digitorum	0.04–0.33	0.2–0.36	Peroneus longus	0.08–0.33	0.24–0.36
Flexor profundus digitorum	0.04–0.33	0.2–0.36	Extensor digitorum brevis	0.05–0.9	0.24–0.36
Flexor brevis pollicis	0.05–0.25	0.2–0.36	Soleus	0.06–0.25	0.24–0.36
Opponens pollicis	0.04–0.28	0.2–0.36	Group 3		
Abductor pollicis	0.04–0.44	0.2–0.36	Gastrocnemius	0.2–0.58	0.44–0.72
Abductor minimi digiti	0.15–0.26	0.2–0.36	Biceps femoris	0.12–0.66	0.44–0.72
Lumbricales	0.05–0.07	0.2–0.36	**C: Trunk**		
Group 4			Pectoralis major	0.05–0.15	0.08–0.16
Extensor carpi ulnaris	0.06–0.65	Upper	Rectus abdominis	0.05–0.18	0.08–0.16
Extensor communis digitorum	0.06–1.0	Excitability	External oblique	0.15–0.2	0.08–0.16
			Trapezius	0.05–0.2	0.08–0.16
Extensor indices	0.2–0.66	0.44–0.72	Teres major	0.05–0.06	0.08–0.16
Extensor longus pollicis	0.05–0.75	0.2–0.36	Latissimus dorsi	0.06–0.16	0.08–0.16
Abductor longus pollicis	0.05–0.52	Lower	**D: Face and Neck**		
			Orbicularis oculi	0.24–0.27	0.48–0.72
Abductor brevis pollicis	0.09–0.62	Excitability	Zygomaticus	0.22–0.44	0.48–0.72
			Quadratus labii superioris	0.32–0.8	0.48–0.72
			Orbicularis oris	0.26–0.32	0.48–0.72
			Frontalis	0.21–0.56	0.48–0.72
			Triangularis	0.08–0.25	0.24–0.36
			Quadratus labii inferioris	0.3–0.036	0.24–0.36
			Mentalis	0.12–0.002	0.23–0.36

From Motorpoint-Charts and Chronaxie Values. TECA Corp., Pleasantville, New York.

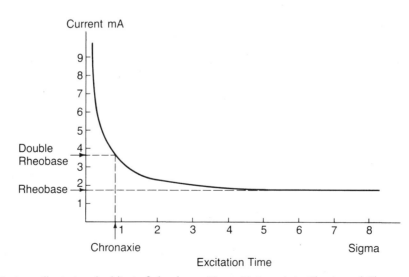

FIG. 8-6. Diagram illustrates doubling of rheobase. (From Motorpoint—Charts and Chronaxie Values. TECA Corp., Pleasantville, New York, with permission.)

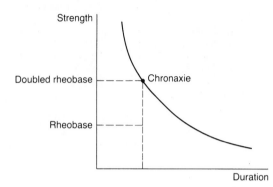

FIG. 8-7. Diagram illustrating relationship between the strength-duration curve and chronaxie.

Comments

Both strength-duration and chronaxie testing procedures may be accomplished on the same specialized equipment, facilitating cross-checking results for accuracy. For all practical purposes, the two tests are exactly the opposite of each other in procedure and purpose.

REFERENCE

1. Wolf S: Electrotherapy. Churchill Livingstone, New York, 1981.

SUGGESTED READINGS

Forster A, Palastanga N: Clayton's Electrotherapy. 8th Ed. Balliere Tindall, London, 1981

Kahn J: Clinical Electrotherapy. 4th Ed. J Kahn, Syosset, NY, 1985

Shriber WJ: A Manual of Electrotherapy. 4th Ed. Lea & Febiger, Philadelphia, 1981

9

Treatment Planning

In this chapter, I "bring it all together." A competent clinician should be able to design an appropriate treatment regimen for any referred condition, provided that adequate data are received from the physician. This information should include a reasonably accurate diagnosis and radiographic and laboratory reports. Knowledge of the patient's prescribed medication is necessary in light of any chemical interactions with either iontophoretic or phonophoretic procedures. The modalities studied here may be administered singly or in varied combinations, since the treatment goals we have are multiple and will surely require several approaches for a comprehensive program. Stress is placed on the treatment of physiopathologic conditions or states, rather than named diseases or syndromes. We are not treating polio, multiple sclerosis, or epicondylitis but are attempting to relieve pain, reduce edema, relax a spasmodic muscle, stimulate a weakened or atonic muscle, minimize inflammation, eliminate a calcific or gouty deposit, increase circulation to an ischemic area, or reduce the congestion in a hyperemic area. Our targets are physiologic and pathologic, not named disease entities, and are at all times patient oriented rather than categorized by classification or standardized regimens as predetermined by policy or practices. The physical therapist must first determine a proper goal for his or her approach, then select those modalities best suited for the target condition. If multiple targets are involved, a spectrum of modalities must be chosen to accomplish all of the goals listed. The specific regimen recommended, the order and frequency of treatment, the parameters needed, and the individual techniques applied will determine not only the success of the treatment administered, but the professional acuity and competence of the clinician.

COMMENTS ON EQUIPMENT SELECTION

You will note that in this chapter electrical stimulation is referred to in terms of

1. Waveform—alternating current or direct current
2. Frequency (Hz) and mode (continuous, surged, or pulsed)
3. Low-voltage, high-voltage direct current, or interferential
4. TENS (transcutaneous electrical nerve stimulation) parameters
5. MES (microampere electrical stimulation) concept

I have avoided using specific brand names, models, or other identifying terminology; electrical stimulation, as far as I am concerned, is generic. Which particular apparatus or procedure is utilized to produce the desired results is the choice of the clinician.

Low-voltage alternating current is utilized for general neuromuscular stimulation when there is no neural pathology present. RD, or reaction of degeneration, present with nerve damage, affects the physiologic function of *alternating current* and negates its use. Pulsed or interrupted *direct current* then becomes the modality of choice. Continuous *direct current* at tolerable intensities will not produce muscle contractions and is used only for iontophoresis.

High-voltage pulsed direct current is selected for deeper penetration. The polar effects are minimal due to the intermittent nature of the waveform. The rule of "the higher the frequency (or the shorter the wavelength) the deeper the penetration" is applicable here. The extremely short pulse duration (in the microsecond range) is much too brief to affect neuropathic tissues, so high-voltage pulsed *direct current* cannot be utilized in the presence of RD. Iontophoresis cannot be administered with this modality since it is not continuous, which is mandatory for ionic transfer.

Interferential current should be chosen when deep penetration is sought with minimal visible/palpable contractions. The medium frequencies found with this equipment are in the 4 to 4.1 kHz range, providing deeper penetration than traditional low-voltage apparatus. However, these frequencies will not produce muscular contractions and are utilized as "carrier waves" to obtain optimal stimulating frequencies (i.e., 80 to 100 Hz) at greater depths by the crossed interferential patterns produced.

TENS is utilized primarily for pain control. It has also been found effective in tissue healing and osteogenic applications with nonunion fractures. Specific and appropriate parameters are listed in Chapter 6.

MES, however, is comparatively new in the field. Here, a subliminal intensity of current is administered. The concept of treating the body with levels akin to the natural parameters, that is, microamperage, is a relatively new concept (see Ch. 5). (Because of the recent development of the MES procedure, it was not included in the original chapter of Treatment Planning.) Microampere-level stimulation may be obtained by parameter adjustment on existing equipment or with the specially designed MES apparatus shown (Fig. 9-1).

To summarize: When electrical stimulation is recommended, suggested, or indicated, the *clini-*cian must determine the appropriate form, mode, and parameters for the individual patient, choosing from among the various applicable modalities available. Manufacturers' claims should not determine the course of treatment or overall management of patients' rehabilitation.

As emphasized previously, iontophoresis may *only* be administered safely and effectively with *continuous direct current.*

Ultrasound is generally administered in the continuous mode. Pulsed ultrasound has no apparent *therapeutic* advantage but will minimize the danger of periosteal burns and other untoward heating effects of the ultrasound.

Cold laser radiation is administered in the continuous mode or at 80 pulses per second for acute conditions. The slower rate of 10 to 20 pulses per second may be utilized in chronic conditions.

In most regimens, I have omitted reference to infrared radiant heat, moist heat packs, cryotherapy (ice), massage, mobilization, traction, and therapeutic exercises, since these modalities are generally administered at the discretion of the attending physical therapist as needed, as part of the individual patient's treatment plan. Therefore, assuming that diagnostic information and data are available to the clinician, I recommend the following approaches to commonly referred conditions.

Computerization

Computers are appearing more frequently on the clinical scene. With electrical stimulation, as well as iontophoresis equipment, preprogrammed

FIG. 9-1. Faceplate of the Myo-matic unit. (Courtesy of Monad Corp., Pomona, California.)

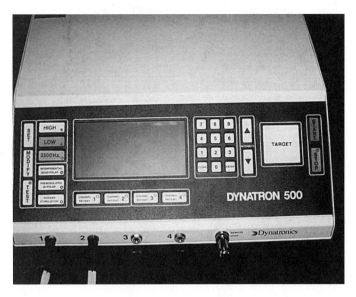

FIG. 9-2. The Dynatron 500 multiple modality electrical stimulation unit, offering high-voltage, low-voltage, medium frequency ("Russian") stimulation and interferential modes in a computerized, programmed, and module-operated chassis. (Courtesy of Dynatronics Corp., Salt Lake City, Utah.)

and programmable units are becoming commonplace in the array of available modalities.

Preprogrammed low-voltage, high-voltage direct current, interferential, "Russian," and MES stimulation procedures are internally programmed by the manufacturers. Parameters are based on favorable clinical experience with each. In most instances, the practitioner may select from a variety of optional sequences (Fig. 9-2). Several of the newer models of TENS units are programmed and programmable (Fig. 9-3).

Programmable units permit the clinician to select, modify, adjust, and administer stimulation automatically. Manual operation is usually an option.

With iontophoresis, the mandatory waveform (continuous direct current) is standard. However, duty cycles, polarity, and intensities may be preprogrammed for automatic operation in addition to the optional manual mode (see Fig. 7-2).

USING ELECTROTHERAPY MODALITIES
Acute Pain

The following modalities are recommended for specific conditions.

1. Lidocaine iontophoresis or phonophoresis
2. Tetanizing alternating current stimulation
3. High-rate (120 Hz) TENS
4. Cold laser

Subacute Pain

1. Hydrocortisone iontophoresis or phonophoresis
2. Magnesium sulfate iontophoresis
3. Salicylate iontophoresis
4. Mecholyl iontophoresis or phonophoresis
5. Tetanizing alternating current stimulation or slow surging alternating current at 100 Hz, if the pain is myogenic
6. Medium rate (40 to 80 Hz) TENS

FIG. 9-3. A new addition to the TENS family is this programmed unit. The patient simply presses controls indicating increased or decreased intensities for preset parameters. (Courtesy of Staodynamics Inc., Longmont, Colorado.)

Chronic Pain

1. Hydrocortisone iontophoresis or phonophoresis
2. Mecholyl iontophoresis or phonophoresis
3. Salicylate iontophoresis or phonophoresis
4. Slow-surged alternating current at 100 Hz
5. Shortwave diathermy
6. Low-rate TENS (1 to 10 Hz)
7. Transarthral surged alternating current if the pain is arthritic
8. Interferential current if the pain is deep

Mild Pain or Discomfort

1. Shortwave diathermy
2. Cold laser
3. Infrared with lidocaine, Iodex, or Myoflex ointments
4. Phonophoresis with any of the above

Inflammation: General or Specific

1. Hydrocortisone iontophoresis
2. Hydrocortisone phonophoresis
3. Salicylate iontophoresis or phonophoresis
4. Cold laser

Ischemia

1. Mecholyl iontophoresis (see Fig. 7-10)
2. Magnesium sulfate iontophoresis
3. Iodex iontophoresis
4. Shortwave diathermy, direct or reflex techniques
5. Mild infrared radiation
6. Surged alternating current stimulation at 100 Hz for ''pumping action''

Edema or Hyperemia

1. Hyaluronidase iontophoresis
2. Mecholyl iontophoresis
3. Salicylate iontophoresis
4. Slow-surged alternating current stimulation at 100 Hz

Calcific Deposits

1. Acetic acid iontophoresis
2. Ultrasound with or without hydrocortisone (phonophoresis)

3. Surged alternating current at 100 Hz to involved muscles (see Fig. 7-12)
4. Acetic acid iontophoresis for myositis ossificans. Providing treatment is begun early in the condition, the sclerolytic effect of the acetate radical on calcium carbonate depositions may be effective in reducing the density and presence of calcific deposit in muscular pathologies, similar to the efficacy with calcific deposits in the bicipital tendon and subacromial bursae.

Neurovascular Paresthesia or Neuropraxia

1. Mecholyl iontophoresis
2. Iodex iontophoresis
3. Cold laser
4. Mild tetanizing alternating or direct current at 100 Hz
5. Slow-surged alternating current at 100 Hz
6. Shortwave diathermy

Myospasm

1. Magnesium sulfate iontophoresis
2. Mecholyl iontophoresis
3. Phonophoresis with hydrocortisone, Myoflex, Iodex, mecholyl, or no chemical
4. Tetanizing stimulation at 100 Hz, followed by:
5. Surged alternating current at 100 Hz
6. Interferential currents for deeper muscle groups

Weak, Atonic, or Atrophied (From Disuse) Musculature

1. Surged alternating current stimulation at 100 Hz if no RD is present
2. Interrupted direct current if RD is present

Gouty Tophi

1. Lithium iontophoresis (see Fig. 7-13)
2. Slow-surged alternating current at 100 Hz

Adhesive Joints or Locking Joints (Snapping Fingers)

1. Calcium chloride iontophoresis
2. Iodex or salicylate iontophoresis
3. Tetanizing alternating current at 100 Hz
4. Surged alternating current at 100 Hz

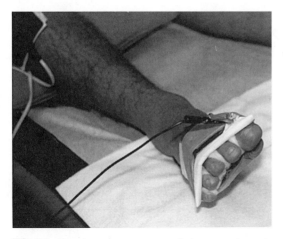

FIG. 9-4. For the management of dermatophytosis (athlete's foot), place cotton balls soaked in 2 percent copper sulfate between toes and overlap with paper towels similarly soaked. Connect to the positive pole electrode, with the negative on the quadriceps.

5. Cold laser
6. Subaqueous ultrasound

Fungi-Dermatophytosis (Athlete's Foot)

Copper sulfate iontophoresis is recommended (Fig. 9-4).

Open Lesions: Healing

1. Minimum erythemal dosage (MED) ultraviolet radiation
2. Zinc oxide iontophoresis or phonophoresis
3. Cold laser

4. Mild electrical stimulation to surrounding tissues with alternating current at 100 Hz or interrupted direct current at very low intensity (i.e., less than 1 mA)
5. As adjunctive therapy to the above clinical modalities, I recommend daily administration of TENS at home to enhance the healing process. Most modern TENS units utilize a biphasic or compensated monophasic waveform that seems to be effective. A suggested technique is to place the electrode astride the lesion so that the current flows *through* the target tissue. A high rate (80 to 120 Hz), medium pulse width (150 μs), and minimal intensities (''just feel''), 1 hour twice daily, is the recommended regimen (Fig. 9-5).

Fracture Healing

TENS at a high rate (120 Hz), with wide pulse width (250 μs) and lowest intensity, is recommended.

Hyperirritability

1. Calcium chloride iontophoresis
2. Mild tetanizing alternating current at 100 Hz or 1,000 Hz

Hypoirritability

Consider the presence of RD, and use interrupted direct current stimulation. The modalities suggested may be administered alone or, more likely, in varied combinations. Most referred conditions involve more than a single symptom or

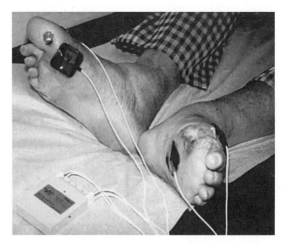

FIG. 9-5. Use of a combination TENS and EMS unit to provide stimulation across open diabetic lesions. (PMS-15, courtesy of Neurologix Corp., Franklin Square, New York.)

treatment goal and require multifaceted approaches. Other modalities not covered in this book are also utilized, along with electrotherapeutic regimens (e.g., hydrotherapy, exercises, massage and mobilization, and splinting).

Low Back

A typical regimen for a low back condition, from the acute stage through the chronic phase, is as follows:

1. When the condition is acute, administer iontophoresis with lidocaine or hydrocortisone to the lower back, including the lumbar nerve roots, followed by tetanizing alternating current across the lower back, at 100 Hz for 5 minutes, and mild infrared radiation with remaining ointment. Give the patient exercise instructions and perhaps mobilization techniques.
2. When the patient's pain decreases, change treatment emphasis to increasing range of motion and mobility. Subsequent treatments should include tetanizing alternating current prior to surged alternating current at 100 Hz, six surges per minute, with a visible contraction of the gluteals at patient tolerance for 10 minutes.

The above example can serve as a generalized approach to the acute- to-chronic transition in musculoskeletal conditions, whether in back, shoulder, elbow, knee, or cervical region. Specific injuries, disease processes, or anatomic complexities are best left to the discretion and expertise of the attending physical therapist at the time of treatment administration. Several particularly interesting entities deserve special mention.

Bell's Palsy

Treat Bell's palsy with hydrocortisone iontophoresis (or phonophoresis). Follow this by interrupted direct current stimulation and, if available, application of cold laser to the 7th cranial nerve distribution-related acupuncture points.

Trigeminal Neuralgia and Tic Doloreux

For trigeminal neuralgia, administer lidocaine or hydrocortisone iontophoresis, cold laser, and high-rate TENS.

Idiopathic Scoliosis

Apply electrical stimulation to the paraspinal musculature at the curve's convexities, usually with a specially designed unit, or use a standard portable electrical stimulation unit, at a rate of from 50 to 120 Hz, medium width of 150 to 200 μs, and at a pulse rate of approximately 3 seconds "on" and 3 seconds "off." This should be used by the patient for 8 hours nightly during sleep.

Sympathetic Reflex Dystrophy

Treat sympathetic reflex dystrophy with mecholyl iontophoresis (Fig. 9-6) and TENS (high rate, medium width, low in density) (Fig. 9-7).

Prostatism

For prostatism, administer shortwave diathermy and electrical stimulation to the sacral nerve roots and gluteal musculature.

Stress Incontinence

For stress incontinence, administer electrical stimulation, alternating current at 100 Hz, six surges per minute for 10 to 15 minutes using an active vaginal electrode, followed by 15 minutes of pelvic diathermy with surface drum or pad electrodes.

Sinusitis, Otitis Media, Bronchitis, Laryngitis

Shortwave diathermy is the treatment for sinusitis, otitis media, bronchitis, and laryngitis, offering immediate and considerable relief for the patient either while other procedures are in progress or as interim treatment. In treatment of laryngitis, ultrasound over the anterior throat has proven an effective technique in rapidly returning voice quality to near-normal (0.5 W/cm^2/1 minute).

Peyronie's Disease

If Peyronie's disease is treated early, hydrocortisone iontophoresis reduces inflammation and discomfort. If the disease is chronic, pain is minimal; however, the anti-inflammatory action of the hydrocortisone has a favorable effect on the fibrotic plaque, often hastening sclerolysis and

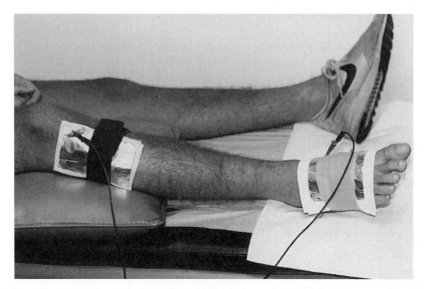

FIG. 9-6. Iontophoresis with mecholyl at the proximal electrode for vasodilation in reflex dystrophy.

improvement. If iontophoresis should prove difficult, phonophoresis may be an alternative method of introducing the 1 percent hydrocortisone ointment. Electrical stimulation, using a standard TENS unit, has also been found to be helpful in reducing localized spasm. The most effective modality to date, however, has been the cold laser for pain control (Figs. 9-8 to 9-11). A comprehensive regimen would consist of cold laser, iontophoresis, and electrical stimulation. If

phonophoresis is utilized instead of the iontophoresis, it should be administered after the electrical stimulation for its antispasmodic characteristics.

Psoriasis

Psoriasis has long been treated with ultraviolet irradiation, using crude coal tar preparations (Goeckermann technique); current procedures substitute plain white petrolatum (Vaseline) for

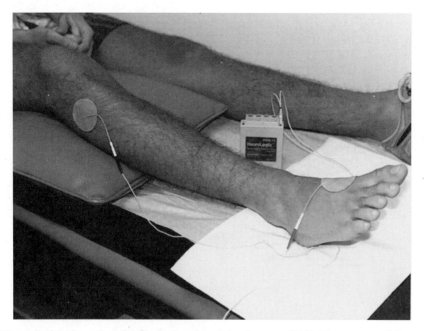

FIG. 9-7. TENS applied (at home) at the fibular head and the dorsum. (PMS-15, courtesy of Neurologix Corp., Franklin Square, New York.)

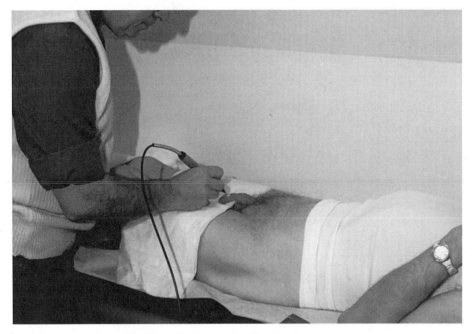

FIG. 9-8. Cold laser to the fibrotic plaque zone.

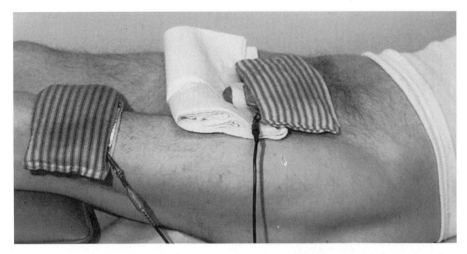

FIG. 9-9. Iontophoresis with 1 percent hydrocortisone ointment, an alternative technique to that shown in Fig. 9-8.

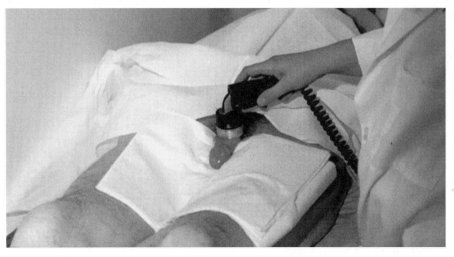

FIG. 9-10. Phonophoresis with 1 percent hydrocortisone ointment.

the tar product with good results. Psoralen plus ultraviolet A (PUVA) techniques are still controversial and should be utilized with care and close cooperation with the attending dermatologist. Resistant sclerotic plaques can often be softened with 1 minute of ultrasound following the ultraviolet radiation. With patches that have been scratched and present an open lesion, I find the cold laser extremely helpful in speeding healing and closure.

Herpes Zoster (Shingles)

The acute pain with intercostal shingles has been modified successfully with cold laser radiation to the appropriate nerve roots. Settings of 80 pulses

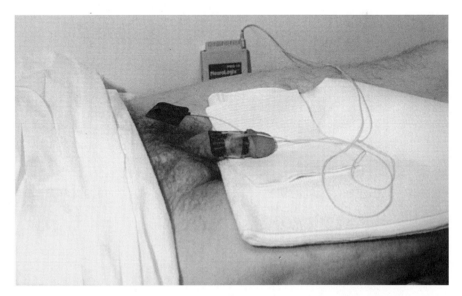

FIG. 9-11. Mild electrical stimulation to a fibrotic area: continuous followed by surged. (PMS-15, courtesy of Neurologix Corp., Franklin Square, New York.)

FIG. 9-12. Cold laser treatment directly to a lipoma.

per second for 30 seconds at each nerve root, daily, are recommended, when practical. Lasing precedes iontophoresis with lidocaine or hydrocortisone.

Lipoma

A single-case study involved a patient with a lipoma on the anterior chin. The attempt to soften or shrink the lesion was only moderately successful. The regimen included cold laser radiation ultrasound and phonophoresis with salicylate. The dimensions of the subcutaneous target were reduced after several treatments (Figs. 9-12 and 9-13).

Sclerosis After Silicone Breast Implant

Personal correspondence with clinicians indicates successful sclerolysis of hardened silicone implant material using standard ultrasound at 0.5 to 1 W/cm^2 for 3 minutes prior to closed capsulotomies.

Pruritus After Dialysis

A single-case study indicated an effective antipruritic procedure for severe generalized itching following dialysis sessions: ultraviolet radiation, MED at 36 inches, total body exposure, twice weekly.

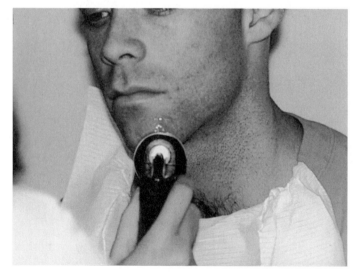

FIG. 9-13. Phonophoresis treatment with iodine and salicylate applied to a lipoma.

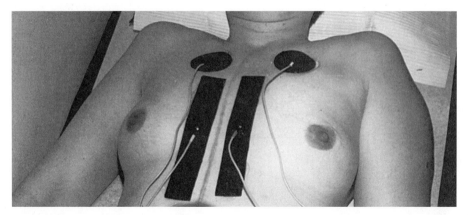

FIG. 9-14. The elongated electrodes are ideally suited to straddle this incision scar. Earlier techniques recommended the additional two proximal electrodes. However, experience indicates their relative unimportance, suggesting omission.

Coronary Bypass Scar Pain

TENS has been found to be effective in controlling postsurgical discomfort. Using the high rate–medium pulse width combination for acute conditions, the scar is straddled by two elongated electrodes. If possible, sterilized electrodes can be placed in position while the patient is still in the operating room. With the connector leads exposed through the dressings, immediate connection is possible in the recovery room. Patients use the TENS continuously, for 24 hours if necessary, and then intermittently as needed (Fig. 9-14).

Erb's Palsy

See Fig. 9-15.

Postcoma Tracheostomy Scar Management

I have found the cold laser to be an effective modality in reducing the constriction of tracheostomy scars. Swallowing improves as the scar softens under the influence of the radiation. Ninety seconds per square centimeter of scar tissue, continuous mode, in direct contact with the target tissues, twice or three times weekly, is the recommended procedure. In addition, electrical stimulation to the usually contracted joints and ultrasound to the shortened tendons complete the total approach to this condition (Figs. 9-16 and 9-17).

The patient in Figs. 9-16 and 9-17 was comatose for 12 months following trauma and complications. Severe sclerosis and adhesions around

FIG. 9-15. In the management of Erb's palsy, early electrical stimulation is advised. It is recommended, however, that the physical therapist's finger be used as the active electrode because of the size of the patient. Here, the reference electrode is placed on the baby's forearm or leg, with the active electrode placed on the physical therapist's dorsum. With transmission gel on the tip of the index finger, pinpoint stimulation is possible with excellent monitoring for intensities.

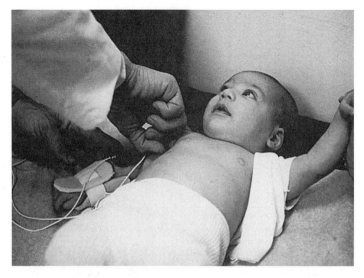

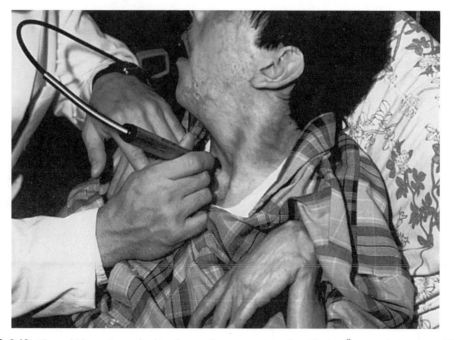

FIG. 9-16. The cold laser is applied to the tracheostomy scar for 90 s/cm^2 several times weekly.

the tracheostomy scar hindered swallowing. The cold laser was administered several times weekly for many weeks with excellent results, leading to softening of the scar, easier swallowing, and improved general functioning. The severe contractions accompanying the tracheostomy problem were also treated with laser and electrical stimulation to the antagonistic musculature, ultrasound for tendon relaxation, with manual stretching, passive range of motion, and massage

to all tissues involved. Although minimal, the results were gratifying.

Cerebrovascular Accident

Since there should be no "RD" with an upper motorneuron dysfunction, ramped alternating current is the ideal choice of current to treat the spastic hemiplegia with cerebrovascular accident. A successful technique is to utilize a tetanic

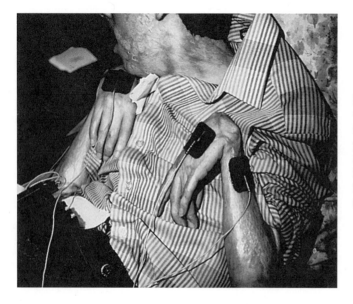

FIG. 9-17. In the patient shown in Fig. 9-16, follow-up electrical stimulation to the over-stretched musculature is applied twice daily at home. A tetanizing mode may be applied to the contracted flexors to offer possible relaxation prior to the surged mode to the antagonists for stretching purposes.

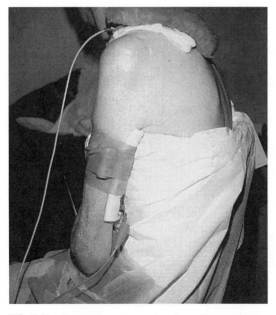

FIG. 9-18. Electrodes shown in place for tetanizing low-voltage alternating current stimulation to the contracted flexor groups.

frequency of 100 Hz for 5 minutes at the origin or insertion point of the deltoid, to patient tolerance, with a visible contraction. Follow this with a surged (ramped) mode, 4 seconds on, 4 seconds off, for 10 minutes. Then, place the electrodes at the origin or insertion along the forearm extensor groups and proceed as before with the 5/10

procedure. If needed, a similar regimen may be followed for the dorsiflexors of the foot, with electrodes at the fibular head and the distal third of the anterolateral aspect of the leg. If contracture into the adducted position of the upper arm is pronounced, precede the deltoid phase with 5 minutes of tetanizing stimulation to the flexor group, with electrodes at the brachial plexus and the *medial* upper arm for the fatiguing effect on the flexors—prior to the extensor stimulation. A similar technique may be attempted with the plantar flexors prior to the dorsi flexor stimulation on the foot (Figs. 9-18 to 9-22).

Dental Application

Cold laser therapy has been found to be an effective form of analgesia with dental and oral surgical pain (Fig. 9-23).

CONCLUSION

It should be noted that in this chapter only successful techniques with individual patients have been reported. No placebo, blind, double-blind, or otherwise controlled research procedures have been involved. Standard equipment and traditional operating procedures were utilized. It is the diagnostic entities treated, the

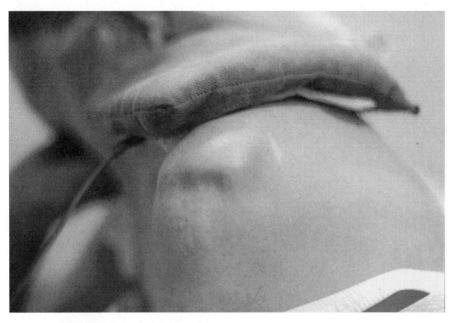

FIG. 9-19. Close-up of the typical subluxation groove at the shoulder.

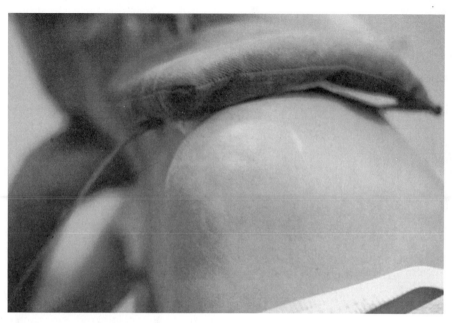

FIG. 9-20. Same location as in Fig. 9-19 under the influence of the stimulation. Note the absence of the groove, indicating actual muscle contraction and not a passive procedure.

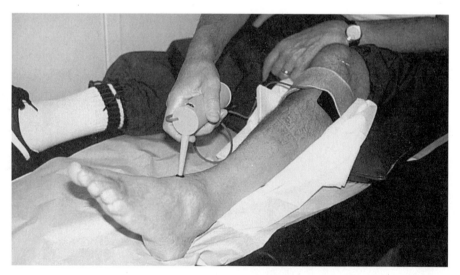

FIG. 9-21. A battery-operated direct current stimulation device for home use. Manual interruption for RD management. (Courtesy of GE Miller Corp., Yonkers, New York.)

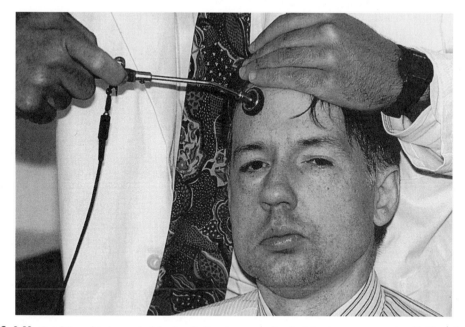

FIG. 9-22. Traditional interrupted (pulsed) direct current for facial nerve neuropathy (Bell's palsy).

FIG. 9-23. Laser shown here being administered intraorally. Dental surgery had been performed on the previous day, with packing in and around the operative site. When the dentist learned of our success with the laser for pain control with the patient's temporomandibular joint problem, he requested the laser for the postsurgical condition. The patient reported complete pain relief for several hours following lasing.

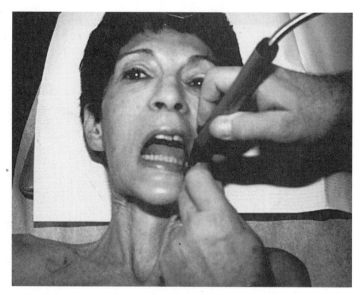

uncommon approaches, and the favorable results that should prompt clinicians' attention and further interest.

There are probably as many "special" techniques as there are practicing physiotherapists; I could not possibly include them *all*. The clinician, using the material herein and all knowledge of anatomy, physiology, pathology, and neuromuscular phenomena at his or her disposal, can design a competent, safe, effective treatment regimen based on astute judgment and scientific rationale.

10

Magnetic Field Therapy

In nontechnical terms, one can say of magnetic field therapy, or magnet therapy, "It works." Magnet therapy is currently receiving a flurry of interest, based upon writings of physicians, scientists, physical therapists, and others with experience with this modality. The scientific bases for magnet therapy are divided between several theories, all of which agree that "it works"—but how?

Magnetic fields present little, if any, danger to the inhabitants of planet Earth, since we are bathed in a constant magnetic flux all our lives. Granted, the amplitude of Earth's natural magnetic field is minimal (in the range of 0.5 gauss [G]); nevertheless, it is there. Geologic evidence of the reversal of Earth's magnetic poles indicates magnetic fields are a changing phenomenon. Some researchers have suggested that events and climatic conditions may be laid to these polar changes; however, this is highly conjectural at present.

Polarity is only one factor of importance with magnetic field therapies. Research has shown that the *negative* (south pole) is primarily sedative, analgesic, and vasodilative. The *positive* (north pole) is stimulative. Even the designation of which pole is which depends upon one's training and scientific background. Geologic and medical terminology regarding polarity differs widely, but once these have been equated, the conclusions are quite similar. The head of a compass points to the Earth's "north pole," which, in effect, makes the compass head a "south pole," since opposite poles attract. On the other hand, if the compass needle is a "true north" pole, the Earth's geographic pole in the Artic would indeed be a "south pole." Confusing, to say the least. A compromise seems to be the term "north-seeking pole" for the compass.

Yet another factor is the apparent differential therapeutic effects of intensity of magnetic flow. The Arndt-Schultz law—which states that a weak stimulus will evoke a favorable response, whereas a strong stimulus may negate a response—applies here, too. Studies have shown that comparatively low gauss intensities are more effective than high intensities: 600- to 3,600-G magnets are the most effective for therapeutic applications. Treatment with much higher intensities, such as 10,000+ G ("the more the better" concept) is definitely ineffective, and perhaps even contrary to our purposes.

Still another factor to be considered: the difference between natural magnetism and electromagnetism. When an electric current is passed through a wire or coil, a magnetic field is generated around the wire. This field can be unipolar if the current is DC and bipolar if AC. Where we seek the advantages of the negative (south) magnetic pole, a bipolar field may not be desired because with a bipolar system the polarity would reverse with the alternating current's frequency.

An interesting phenomenon is the physiologic preference for a nonstatic field, that is, a field that pulses. This is easily accomplished with man-made electromagnetic fields that have added parameter selection controls.

However, the greater efficiency with natural magnet therapy—with lodestone, magnetite, and other magnetic materials—is thought to be brought about by the patient's own movements while wearing the magnets, compared with the use of electromagnetic fields, which necessitates the patient's presence in an office facility for a period of time for a "treatment." Normal movements during daily activities provide optimal changes in the field by altering the shape, conformity, and absorption qualities of the target tissues. (Recall that magnetic fields penetrate universally. Earth's magnetic field cannot be "turned off.")

The actual clinical application of natural magnets is mainly a matter of experience, individual techniques, and use of recommended procedures. Large units (e.g., 4 × 6 inches [10 × 15 cm]) are suggested for treatment of low back discomfort; disks about 2 inches (5 cm) in diameter serve well for the shoulder, elbow, knee, hand, and foot. Held in place with Velcro straps or adhesive tape, these units are unobtrusive, easily removed if necessary, and relatively inexpensive. In addition, there are small "dots," about the size of a dime (1.5 cm), that are extremely effective when placed either on the point of pain or at appropriate acupuncture points. Catalogs from the manufacturers or dealers will indicate the pricing. Because magnets are *not* considered "medical devices," neither they nor required prescriptions are covered by insurance carriers for reimbursement. These units may be worn for hours during normal business and athletic activities. Also available, with magnets incorporated, are car seats, cushions, mattresses, inner soles, and even orthotic devices. Contraindications are few: avoid use in the presence of pacemakers, pregnancy, possibility of hemorrhage, or near transdermal medical patches.

Just how magnets work is still not clear. Various researchers have postulated that magnetic fields are "read" by the body as electrical stimulation, producing endorphin release; others cite the gate theory of differential neural transmission alterations or realignment of molecular structures, effecting changes in enzyme and ionization products (the Hall effect, i.e., the production of heat when ions are in a magnetic field).

Every tissue in the body is bathed in blood. The main ingredient of blood is hemoglobin, with the "heme" representing magnetic iron. Surely there is a general effect on *all* tissues in achieving polar alignment (i.e., north to south) when in a magnetic field. Optimal alignment may very well be involved in the increased blood flow.

Regardless of the how, the fact remains that *it does work*. Researchers must delineate the actual systems by which magnetic fields join the other electrotherapeutic modalities in the battle against pain.

There are many physiologic benefits attributed to magnetic fields, including increased oxygenation, enhanced blood flow, and hormonal production. It is suggested that the body "reads" magnetic fields as an electric phenomenon (i.e., stimulation), especially if it is a pulsed flow. The pulsing of electromagnetic devices is programmed for multiple frequencies. When a patient wears natural magnets, however, the flux is constant, yet varied by normal body movements that alter and modulate the effective magnetic fields, simulating mechanical pulsing. From past experience, we know the body apparently responds better to intermittent (pulsing) stimulation than to continuous forms.

TECHNICAL DATA[1]

Normal cell potential is about 90 mV. In the presence of inflammation this rises to about 120 mV; with degeneration, it falls to 30 mV. This points up the phenomenon described by Dr. Bjorn Nordenstrom[2]: The increased electrical resistance around injured tissues prevents the low potential capillary ionic flow from entering the zone and enhancing healing.

Magnetic fields, however, permeate all cells, regardless of potentials. Magnetic fields are not absorbed. A concentrated "bundle" of magnetic lines is termed *paramagnetism;* a decentralization of magnetic lines is termed *diamagnetism.* The human body is considered to be neutral in relation to both paramagnetism and diamagnetism. Therefore, magnetic field lines permeate all parts of the human body, differing from standard forms of therapy.

Normal cells have a basic electrical potential, which is necessary for them to function. Cells that lack this potential are nonviable. Diseased and damaged cells have altered rest potentials. Pulsating magnetic fields influence ions around the injured zone. The rest potential is highly influenced by the magnetic field. In turn, ionic exchange at the cellular level influences oxygen utilization, needed for healing. Pulsed magnetic fields also reduce pain by increasing oxygen pressure and perfusion, as well as capillary blood flow, reducing the accumulation of metabolites in the area.

Pulsed magnetic fields do *not* produce heat. Patients with implants, other than pacemakers, may be treated. Fractures may be treated with casts, since magnetic fields permeate all materials. Indications include the treatment of orthopedic injuries, rheumatism, wounds, burns, circulatory diseases, and neurologic disorders. Effects may last 6 to 8 hours.

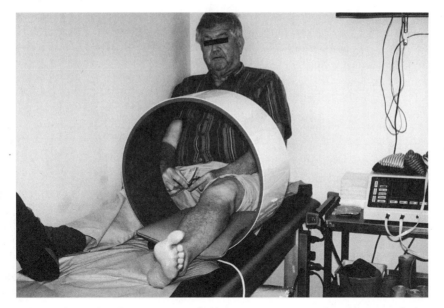

FIG. 10-1. The "loop" device for electromagnetic field therapy, in position for treatment of a knee injury.

Much as I recommend TENS, I recommend the application of magnets for pain control between office visits, when the patient is traveling, or during periods when formal treatments are not practical. I have found magnets to be 90 percent effective in controlling pain, with few patients reporting they received no favorable results. In most instances, partial to full pain relief is obtained when the patient complies with my recommended procedures. Generally, I suggest wearing the magnets for 3 hours in the morning, then again for 3 hours late in the afternoon or evening.

Occasionally, patients wear them from morning until bedtime for continuous relief. Many athletes wear magnets during their events, with satisfactory results.

My original exposure to magnetic therapy was with a device of a Canadian manufacture, the Centurion.[3] Large circular "coils" into which the patient placed a limb or the torso were connected to an electromagnetic generator (Fig. 10-1). The resulting magnetic field permeated the coil, surrounding the affected anatomy with a strong magnetic field. The coil strength was dependent

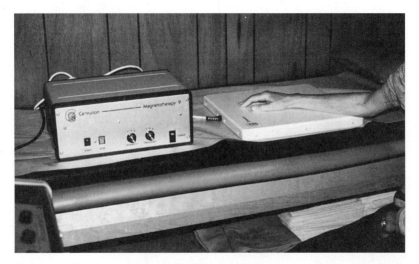

FIG. 10-2. Flat-plate device for electromagnetic field therapy for treatment of an arm and wrist condition.

upon the operating frequencies and ranged from 20 to 50 G, pulsed at 2, 15, or 30 Hz. Treatment time was for 30 minutes; there was no sensation for the patient. It was necessary to place a small, handheld magnet in the field to detect the pulsing of the field. I eventually obtained a flat-plate device in place of the awkward circular coil (Fig. 10-2). Patients could lean back against this plate, place a limb on the plate, or even sit on it. The magnetic field was operational only when the generator was "on," differentiating electromagnetic fields from natural magnetic fields, obtained from such materials as lodestone and magnetite. Results were consistently good.

I was later introduced to the small, natural magnets for personal patient use.[4] Unipolar, "tectonic" magnets provide the negative phase recommended for pain relief. Various sizes are available to be worn by patients; these are held in position with Velcro straps or adhesive, as described earlier. They are available in inner soles, back supports, automobile seats, mattresses, and a wide selection of shapes for various anatomic areas (Figs. 10-3 through 10-5). These personal magnets range in strength from 1,100 thru 12,000 G. Unlike electromagnetic devices, they are always "on."

I have had *no* reports of unwanted side effects from any patients, other than a man who made the mistake of placing his magnet in the same area as his credit cards—which were wiped clean. The same caution is advised regarding use near computer software and screens, video and music tapes, pacemakers, and in pregnant patients

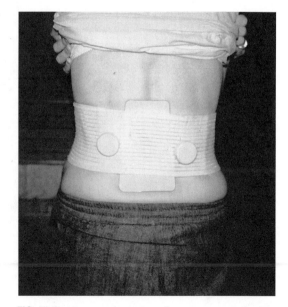

FIG. 10-3. Magnet placement for treatment of a low back condition. The central unit is flanked by two smaller discs for additional field strength.

since the effect of magnets on a fetus is unknown.

Several manufacturers now offer magnets over the counter for consumer convenience. A careful check of the gauss rating is advised, and the cautions listed on the product should be followed. Both as an office modality and a home treatment device, magnet therapy is proving to be an effective approach, offering nonnarcotic, safe, and reasonably priced pain control.

FIG. 10-4. Domino-shaped magnet for treating discomfort associated with carpal tunnel syndrome. The magnet is held in place with a Velcro strap.

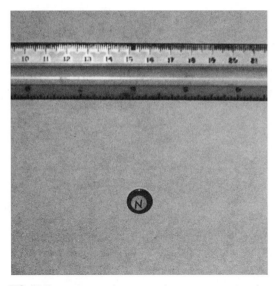

FIG. 10-5. Dime-sized (1.5 cm diameter) magnet for pinpoint sites of pain. The ''N'' on the magnet indicates the negative pole.

REFERENCES AND RESOURCES

1. Centurion Home System Operational Manual, Calgary, Alberta, Canada
2. Nordenstrom B: Biologically Closed Electrical Circuits. Sweden, 1983. In Taubes G: An electrifying possibility, Discover, April 1986, p. 23
3. Centurion Home System
   Elec Western Medical Devices, Ltd
   1015 Matheson Blvd, Suite 8
   Mississauga, Ontario, Canada L4W3A4
   416-238-4860
4. Magnetherapy, Inc.
   760 U.S. Highway #1
   N. Palm Beach, FL 33408
   800-625-9736

SUGGESTED READINGS

Becker RD: Cross Currents. Tarcher Putnam, New York, 1990

Hallett M, Cohen LG: Magnetism. JAMA 262:538, 1989

Kahn J: Magnets anyone? PT/OT Today, August 26, 1996, p. 9

Kahn J: Which way is up? PT Advance, February 26, 1996, p. 4

Trock DH et al: A double-blind trial of the clinical effects of pulsed electromagnetic fields in osteoarthritis. J Rheumatol Fall 1992.

*Washnis GJ, Hricak RZ: Discovery of Magnetic Health. Nova Publishing Co., Rockville, MD, 1993

*Whitaker J, Adderly B: The Pain Relief Breakthrough. Little, Brown & Co., Boston, MA, 1998

Zimmerman J, Hinrichs D: Magnetotherapy: An Introduction. Bio-Electromagnetic Institute of Reno, NV, Physical Therapy Products, March 1955, p. 22

*Highly recommended.

11

Recent Developments in Electrotherapy

COMBINED MODALITIES

Simultaneous electrical stimulation and ultrasound has yet to be proven more effective than each modality applied separately. A stimulation treatment lasts considerably longer than an ultrasound treatment. To avoid overdosing with ultrasound, it seems appropriate to utilize the "pulsed" mode of ultrasound, which, in effect, reduces the "on" time. This is not an insurmountable problem for clinicians who choose to use the "continuous" mode of ultrasound for maximum effect. Reduction of dosage is a simple answer, if practical for the intended target tissues.

My main concern with this combined treatment, however, lies in another area: Electrical stimulation is designed primarily to obtain muscular *contraction*. One of the characteristics of ultrasound is to increase tendon *extensibility*. It seems improbable that one would exact maximal contraction *or* maximal relaxation (extension) if both modalities were administered simultaneously. If one were to assume that the relaxation obtained with the ultrasound would enhance the contraction with the electrical stimulation, the time parameters for each would necessarily have to be ignored. My preference, obviously, is to utilize both modalities separately for maximum effect with *each*.

An ultrasound treatment lasts between 2 and 8 minutes, while an electrical stimulation session might be considerably longer (i.e., 10 to 20 minutes). Why not administer the stimulation for the proper duration, follow with ultrasound afterward, and thus obtain the full benefit of each? The combination *does* save time and is reported to facilitate locating "trigger points." However, the few minutes saved seem insignificant when matched against treatment efficacy. Furthermore, I feel any competent physical therapist in practice more than 6 months should have little difficulty in locating trigger points manually.

The marriage of electrical stimulation with cold laser radiation is another combination worth commenting on. The leading manufacturer of the HeNe cold laser offers a model that includes microampere stimulation separately or combined with the laser. Variable waveforms make this unit useful in the treatment of several conditions with simplified application techniques, using the laser wand for both modalities. The administration of both separately offers physical therapists ample opportunities to evaluate the efficacy of either. Just how necessary or important the combination is remains to be seen.

ELECTROSTIMULATION IN HIV INFECTION

A new application of electrical stimulation is being considered in the battle against human immunodeficiency virus (HIV). Viruses are reported to lose the ability to reproduce after exposure to microampere stimulation. The enzyme necessary for reproduction is curtailed. Its loss also prevents white blood cell clumping, a sure sign of viral infections.[1]

REFERENCE

1. Lyman WD et al: Electrostimulation in HIV infection. Cited in Science News, as reported in Phys Ther Adv, August 3, 1992

SUGGESTED READINGS

Rachlin ES: Myofascial Pain and Fibromyalgia. CV Mosby, St. Louis, 1944

Low J, Reed A: Electrotherapy Explained. Heinemann Medical, LD, London, 1990

Kahn J: Challenges or Deterrents? PT/OT Today March 25, 1996, p. 5

12

General Considerations

It is necessary to comment on the billing and management problems of physical therapy under current regulations, especially Medicare.

Many of the procedures advocated in this manual are not "standard" in most third-party payor schedules, whose consultants, in most cases, are *not* physical therapists and have little, if any, knowledge or experience at the hands-on level of physical therapy care available today.

Clinicians should avoid listing diagnoses in general terms and concentrate on the actual pathologic conditions treated. Examples are "painful muscles or joints" instead of "osteo-arthritis"; "weak muscle" rather than "stress incontinence." Many of the conditions we treat are considered by schedules to be surgical, medical, or, worse, "incurable." Therefore, these conditions would *not* appear on the payor's clerk's list of entities covered by physical therapy. Consequently, I recommend that "conditions treated" be listed as "pain," "spasm," "strain/sprain," "inflammation," "weakness," "dysfunction," and so forth, pertaining to the anatomic target in question, as listed in the code books. (I have often listed "soft tissue dysfunction" as a catch-all phrase for unusual conditions.)

Too often, physical therapy is considered to be limited to orthopedic conditions only, leading to denial of payment when nonorthopedic conditions are involved. I advise clinicians to be adamant regarding these matters and to present documented arguments and evidence if and when challenged. Even in established territories in the physical therapy "ballpark," innovative and nontraditional procedures and techniques can be subject to denials. *Traditional,* of course, refers to treatments identical to those offered "by everyone else for the past 75 years" and is "familiar" to the consultant, regardless of its efficacy.

It should further be noted that in many states, including my own, New York, physical therapists practice upon physician referral. Denials, therefore, pit the consultant against the referring physician. It is interesting to learn that the physician's payment is *not* denied in these instances.

Electrotherapy is particularly problematic to nonphysical therapy consultants. Their exposure is often limited to the information provided by manufacturers' sales representatives, which in some cases is only what they have observed in clinics, departments, and limited facilities. The selection of modalities, procedures, and especially *techniques* now available for patient care is considerably broader than in the past.[1]

Another problem frequently reported by my students out on affiliation or at their first positions is the reluctance on the part of attending physicians to approve of suggested techniques because of the physicians' unfamiliarity with *our* standard procedures, all of which fall under the aegis of our statutory physical therapy programs curricula. A demand by the physician for articles on the topic in question is usually followed by a rejection as presented *unless authorized by physicians.* Most of the practice-oriented references in electrotherapy are written by physical therapists, and thus the problem becomes more complex and frustrating. Although there *are* physician-authored articles, few, if any, relate directly to the "hands-on" level phase of management.

Electrotherapy is *not* included in medical school curriculum. It *is,* however, a regular mandatory phase of study in the physical therapy curriculum and is covered on the licensing examination. Just as a physician would not tell a pharmacist how to compound, a podiatrist how to practice, a dentist how to extract a tooth, an optometrist how to refract, why, then, should someone *not* trained in our discipline be reluctant to accept suggestions, recommendations, and procedures from those who *are* trained, tested, and professionally licensed? (Yet athletic trainers, in many instances, may administer treatment, including physical therapy or electrotherapy, without medical supervision or referral.)

It is to be hoped that in the future there will be a greater reliance on the clinical judgment of professional physical therapists when it comes to direct patient care. This reliance should be based on treatment outcomes, documentation, and considerable regard for the favorable reports from the rest of the medical world.

REFERENCE

1. Kahn J: Keep the physics in physical therapy. Phys Ther Forum, May 1, 1992, p. 44

Equipment Selection

When the time comes for you to purchase or recommend purchase of electrotherapy equipment, several points should be kept in mind:

1. Bigger is not necessarily better. Home care situations, small office space, and limited hospital space suggest the selection of small, portable units whenever practical. Who is going to push this unit around?
2. Do not count on any single unit to offer all of the modalities needed. This becomes a serious problem if and when repairs are necessary.
3. Many companies offer rental prior to, or instead of, purchase. This is a useful concept if you are not sure of the need for the particular piece of apparatus.
4. Combined units (e.g., ultrasound and electrical stimulation) may not be suitable in facilities where mobility and multiple patient schedules prevail. Separate units can be utilized in separate areas simultaneously. Avoid treating two patients with a single device.
5. Simplicity in parameter controls is better than complexity. Why pay for a washing machine with 20 different cycles when you are only going to use "wash, rinse, and spin"? You will find many of the staff never touch or change the settings established on the day of delivery.
6. If you must put out low bid orders, make certain you request the specific parameters with desired equipment or you may wind up with an "equivalent" (i.e., the least expensive model).
7. Make certain the dealer demonstrates the unit prior to purchase and guarantees service, maintenance, and, whenever possible, use of loaner equipment during "downtime" on the purchased unit. In-service training is always advantageous, especially with a large staff and students.
8. Make sure operational manuals come with the units.
9. Check serial numbers on delivered models as well as current inspection certification.
10. I also recommend taking a photograph of the unit for office records and future dealings.
11. Having spares on hand for battery-operated units is a must.
12. Attending conference and convention exhibit halls is a far better way to learn of the newer equipment than through catalogs. See, try, and feel before ordering. I recommend that the entire staff know what the unit "feels" like before administering treatments to patients.

Questions Most Often Asked

1. Why do I get burns with iontophoresis?

Most iontophoresis burns are due to poor technique and not the concept: Poor skin-electrode interfaces; intensity too high; Velcro straps too tight; and electrodes too small, too dry, and not enough size differential between anode and cathode. Wrong polarity often is a culprit. Using other than continuous direct current is a common error.

2. Why do pharmacists give me such a hard time with iontophoresis chemicals?

Pharmacists are not trained in iontophoresis and often will give a fledging physical therapist a tough time. They think in terms of compounding the ionic sources, despite the availability of excellent over-the-counter preparations. "You can't ionize that substance," or worse, is an often-heard rebuke.

3. Can I use ultrasound over metallic implants?

Why use ultrasound over a metallic implant in the first place? The interface between tissues and the steel begs for a collection of heat buildup. Reflected soundwaves from the metallic face could be a problem, too. For pain control, there are other methods less problematic. In some of the older prostheses, danger of damage to the cement matrix is present. Iontophoresis penetrates less than 1 mm in depth, electrical stimulation slightly more, and microamperage stimulation's target is the circulatory circuit. If ultrasound is the *only* modality you have (as is often the case), then I would opt for a loaner from a local dealer of another modality for this patient.

4. Can I administer iontophoresis with high-voltage direct current equipment?

The only waveform iontophoresis is continuous direct current! The extremely short pulse duration of high-voltage apparatus, combined with the microamperage produced, is insufficient to introduce complex ions into the tissues in any appreciable quantity. With the formula for ionic transfer (see p. 120), you would have to leave the electrodes in place for hours or days.

5. Can I use phonophoresis under water?

The soundwave front as it emerges from the transducer head hits the water, and its power and dispersion are immediately compromised. Any chemical would be dissociated by the water and lose its concentration. If you were foolish enough to use the whirlpool tank for this purpose, the situation would be compounded by the gallons of water involved instead of the smaller amount in a plastic or fiberglass pan. The whirlpool should not be used for electrical or ultrasound treatments.

6. Is it true that ultrasound therapy must be limited to 20 treatments?

I have never found anything in the literature to establish the 20-treatment limit with ultrasound. Some of the manufacturers' manuals and recommendation charts suggest a given number of treatments for specific conditions; however, there is little, if any, reference to this limitation in the clinical literature. If you have not obtained satisfactory results in 20 treatments, why continue? And, if you are gaining good ground at 20, why stop?

7. Do I need special equipment for vaginal electrical stimulation for stress incontinence?

The only special equipment needed for vaginal stimulation with stress incontinence is the vaginal electrode. This is available commercially. The EMPI Corp. now offers a specialized electrode/stimulation system for this purpose.

8. Why will physicians not approve TENS for nonunited fractures?

TENS for nonunion fractures is relatively unknown to physicians. They have been oriented to use a more complex, expensive system. Most of the successful trials with TENS have been accomplished by physical therapists who, in turn, amaze the referring physician with the technique.

9. Is the laser experimental?

The helium-neon, 632.8-nm cold laser is *not* experimental. It is under *clinical investigational status*. Those willing to complete the necessary forms, and toe the protocol lines set up by the FDA, may use this fascinating, exciting, and effective modality. It is worth all the trouble.

10. Can I use vinegar instead of acetic acid for calcium deposits?

No. Acetic acid is what gives vinegar its smell and taste. Vinegar is an organic substance and, as such, is not readily ionized. Use the real stuff—2 percent acetic acid.

11. Are iontophoresis and phonophoresis interchangeable clinically?

No, iontophoresis and phonophoresis are not equivalent. *Ions* are introduced with iontophoresis, while *molecules* are introduced by the ultrasound waves. There is a world of difference between *ionic sodium* and *ionic chlorine*—and *molecular table salt!* Furthermore, since soundwaves are not electrical in nature, no ionization takes place.

12. Is it true that diathermy is no longer permitted to be used?

No. Shortwave diathermy is still a viable modality used generally all over the world, but very seldom in the United States. We are limited to 3 wavelengths for medical purposes (7, 11, and 22 m), which are more than sufficient for our purposes. Unfortunately, the topic is rarely taught, nor are students exposed to this useful modality by their instructors, for whatever reasons you can imagine.

Note: *Principles and Practice of Electrotherapy* is now available in Spanish and Polish. For details write to the International Rights Department, W.B. Saunders Company, The Curtis Center, Independence Square West, Philadelphia, PA 19106-3399.

Index

Note: Page numbers in *italics* refer to illustrations; page numbers followed by t refer to tables.